THE HUMAN MIND'S IMAGININGS

The Human Mind's Imaginings

*Conflict and Achievement in
Shelley's Poetry*

MICHAEL O'NEILL

CLARENDON PRESS · OXFORD
1989

Oxford University Press, Walton Street, Oxford OX2 6DP

Oxford New York Toronto
Delhi Bombay Calcutta Madras Karachi
Petaling Jaya Singapore Hong Kong Tokyo
Nairobi Dar es Salaam Cape Town
Melbourne Auckland
and associated companies in
Berlin Ibadan

Oxford is a trade mark of Oxford University Press

Published in the United States
by Oxford University Press, New York

British Library Cataloguing in Publication Data
O'Neill, Michael
The human mind's imaginings: conflict and
achievement in Shelley's poetry
1. Poetry in English. Shelley, Percy Bysshe,
1792–1822. Critical studies
I. Title
821'.7
ISBN 0-19-811748-5

Library of Congress Cataloging in Publication Data
O'Neill, Michael (Michael S. C.)
The human mind's imaginings.
Bibliography: p.   Includes index.
1. Shelley, Percy Bysshe, 1792–1822—Criticism
and interpretation.   I. Title.
PR5438.05 1989 821'.7 88-34507
ISBN 0-19-811748-5

Phototypeset by Dobbie Typesetting Limited
Printed in Great Britain

To my parents

Acknowledgements

THIS book began life as a thesis. I am grateful to my supervisors, John Buxton and John Kelly, and to the examiners of the thesis, John Bayley and the late Geoffrey Matthews, for their advice. I would like to thank the University of Durham for granting me a term of sabbatical leave during which I have been able to complete the book. I am indebted to many people for stimulus, encouragement, and criticism, especially David Fuller, David Hartnett, John Kerrigan, Jamie McKendrick, Posy O'Neill, Gareth Reeves, Stephen Wall, Kim Scott Walwyn, J. R. Watson, Timothy Webb, and Jonathan Wordsworth; they have helped to prevent the book from turning into what Shelley calls 'a Sort of Asymptote which seems ever to approach & never to arrive' (*Letters*, ii. 179). Needless to say, responsibility for any shortcomings is my own.

A slightly different version of Chapter Six appeared in the *Keats-Shelley Review*, ii (1987), 105–33. Chapter Seven is more or less reprinted from *Essays in Criticism*, xxxvii (1987), 135–57. Chapter Eight is a revised version of an essay included in J. R. Watson (ed.), *An Infinite Complexity: Essays in Romanticism* (Edinburgh, 1983), 161–80. I am grateful to the editors of *Essays in Criticism*, to the editor of the *Keats-Shelley Review*, and to the Publications Board of the University of Durham for permission to reprint.

Contents

Note on Texts

UNLESS stated otherwise, quotations from and references to Shelley's poetry and prose are taken from *Shelley's Poetry and Prose*, ed. Donald H. Reiman and Sharon B. Powers (see Abbreviations and the Bibliography for full details). I consciously depart from this edition in two places: I substitute a comma for a full stop at the end of line 25 of *Alastor* and emend ' multidinous' to 'multitudinous' in *Prometheus Unbound*, IV. 253. Unless stated otherwise, the texts used for quotations from and references to the poetry of Blake, Browning, Byron, Coleridge, Keats, Milton, Spenser, and Wordsworth are the Oxford Standard Authors editions (see the Bibliography for full details). Details of the texts used for quotations from the work of Crabbe, Shakespeare, Stevens, and Yeats can be found in the Bibliography.

Abbreviations

LISTED below are abbreviations used in the text and the notes. Bibliographical details always refer to the particular edition I have used; where I have referred to a reprint of a particular edition I give first the date of the edition, then a semicolon, then the place of publication and date of the reprint.

Baker
: Carlos Baker, *Shelley's Major Poetry: The Fabric of a Vision* (1948; Princeton, NJ, 1973).

Cameron
: Kenneth Neill Cameron, *Shelley: The Golden Years* (Cambridge, Mass., 1974).

Chernaik
: Judith Chernaik, *The Lyrics of Shelley* (Cleveland, Ohio, and London, 1972).

Cronin
: Richard Cronin, *Shelley's Poetic Thoughts* (London, 1981).

Davie
: Donald Davie, *Purity of Diction in English Verse* (enlarged edn., 1967; London, 1969).

DC
: Harold Bloom, Paul de Man, Jacques Derrida, Geoffrey H. Hartman, J. Hillis Miller, *Deconstruction and Criticism* (London and Henley, 1979).

EIC
: *Essays in Criticism.*

ELH
: *English Literary History.*

EOS
: Miriam Allott (ed.), *Essays on Shelley* (Liverpool, 1982).

Julian
: *The Complete Works of Percy Bysshe Shelley*, The Julian Edition, ed. Roger Ingpen and Walter E. Peck (1926–30; 10 vols., London, 1965).

Keach
: William Keach, *Shelley's Style* (New York and London, 1984).

KSJ
: *Keats–Shelley Journal.*

KSMB
: *Keats–Shelley Memorial Bulletin.*

Leavis
: F. R. Leavis, *Revaluation: Tradition and Development in English Poetry* (1936; Harmondsworth, 1972).

Leighton   Angela Leighton, *Shelley and the Sublime: An Interpretation of the Major Poems* (Cambridge, 1984).

Letters   *The Letters of Percy Bysshe Shelley*, ed. Frederick L. Jones (2 vols., Oxford, 1964).

PMLA   *Publications of the Modern Language Association of America.*

PP   *Shelley's Poetry and Prose*, Norton Critical Edition, ed. Donald H. Reiman and Sharon B. Powers (New York and London, 1977).

PW   *Shelley: Poetical Works*, Oxford Standard Authors, ed. Thomas Hutchinson, corr. G. M. Matthews (2nd edn., London, New York, and Toronto, 1970).

SIR   *Studies in Romanticism.*

SM   Harold Bloom, *Shelley's Mythmaking* (1959; Ithaca, NY, 1969).

Wasserman   Earl R. Wasserman, *Shelley: A Critical Reading* (Baltimore and London, 1971).

Zillman   *Shelley's 'Prometheus Unbound': A Variorum Edition*, ed. Lawrence John Zillman (Seattle, 1959).

Introduction

The Human Mind's Imaginings: Conflict and Achievement in Shelley's Poetry—my title borrows a phrase from the end of 'Mont Blanc' (143) that serves to highlight major emphases of this study. In it I contend that Shelley's best poetry deals profoundly if at times obliquely with what it means to be 'human'; that the 'mind', consciousness, occupies a central place in the poems, establishing subtle relationships with what lies beyond, within, or below itself; and that these relationships are explored through the poetry's 'imaginings', verbal inventions that are frequently aware of themselves as fictional constructions, even as they refuse to give priority to non-poetic forms of knowing. The very word 'imaginings' is, by contrast with 'imagination', alive with a sense of provisionality. But the key words in the title are 'conflict' and 'achievement'; put briefly, my central argument is that the exploration of conflict lies at the heart of Shelley's poetic achievement.

Despite much fine work on Shelley in recent years, there is still scope for a book that grapples with the issue of the poetry's worth. In 1976 Thomas McFarland remarked that 'modern Shelley scholars . . . simply ignore the fact that Hazlitt, Arnold, and Leavis, that is, three of the half dozen or so greatest critics of English literature since Shelley's time, all call into severest question Shelley's poetic quality and importance'.[1] The judgement is too summary; for one thing, it fails to take account of Harold Bloom's *Shelley's Mythmaking*, still the most important because most critically discriminating study of the poetry. With acidic gusto Hazlitt had mocked the poet's style as 'a passionate dream, a straining after impossibilities, a record of fond conjectures, a confused embodying of vague abstractions'.[2] Leavis, more in revulsion than revaluation, attacked a 'notable lack of self-knowledge and a capacity for ecstatic idealizing'.[3] But, as

---

[1] 'Recent Studies in the Nineteenth Century', *Studies in English Literature: 1500–1900*, xvi (1976), 694.
[2] From a review of Shelley's *Posthumous Poems* (1824) reprinted in Theodore Redpath, *The Young Romantics and Critical Opinion, 1807–1824: Poetry of Byron, Shelley, and Keats as Seen by their Contemporary Critics* (London, 1973), 388.
[3] Leavis, 208.

Denis Donoghue points out, Bloom revitalized the debate about Shelley's supposed immaturity by seeing the poetry as dramatizing, in Donoghue's phrase, 'the endlessness of desire'.[4] His book addresses fundamental issues raised by Shelley's poems: he argues forcibly and cogently against allegorical accounts of the poetry that evade its imaginative challenge;[5] he calls into question the relevance to the poet's work of the new critical dogma of the 'universal necessity of concrete imagery';[6] and he is sensitive to the presence within Shelley's best poems of ironic awareness and undertows of feeling.[7] However, my admiration for *Shelley's Mythmaking* is tempered. Bloom's arguments are in places over-insistently propelled by a determination to read the poetry as concerned with 'myth: the process of its making, and the inevitability of its defeat'.[8]

As well as ignoring Bloom, McFarland's remark pre-dates thoughtful defences of the poetry offered by Richard Cronin (in *Shelley's Poetic Thoughts*) and William Keach (in *Shelley's Style*), among others. My debt to and dissent from the views of Bloom, Cronin, and Keach will be apparent in ensuing pages. But McFarland is alert to a tendency shown in much recent criticism to take for granted Shelley's status as a great poet. The need for a sympathetic yet discriminating reading of the poetry is still evident. An attempt to offer such a reading, this book focuses in detail on most of Shelley's major works. It is at once exploratory and evaluative; its structures of argument are bound up with and emerge from the process of critical analysis.

Only while analysing a 'particular poem' can one act on William Empson's excellent advice: 'you must rely on each particular poem to show you the way in which it is trying to be good'.[9] In the same paragraph Empson expresses his reluctance to write about poems that fail. Yet his phrasing—'trying to be good'—allows for the possibility of failure; it also encourages one to keep an open mind about the reasons for a poem's success. A poem such as *Prometheus*

---

[4] 'Keach and Shelley', *London Review of Books*, 19 Sept. 1985, 12.
[5] See e.g., his tart comment in the course of a reading of *Prometheus Unbound*, Act IV: 'The researches of Sir Humphry Davy will not aid in the difficult labor of understanding what it is that Ione *sees*', SM 141.
[6] SM 39.
[7] See his discussion of 'the too-little regarded and quite formidable irony of Shelley', *SM* 94.
[8] SM 8.
[9] *Seven Types of Ambiguity* (3rd edn., 1961; Harmondsworth, 1973), 26.

Unbound may try to be good in one way (the construction of a myth which will embody 'beautiful idealisms of moral excellence', *PP* 135). But analysis may suggest it is also trying, at an important level of its being, to be good in another (the recognition of the problems involved in constructing such a myth). As a result I view with intermittent scepticism and caution the notion that the poetry should be seen as (and praised for) fulfilling Shelley's intentions or aesthetic views as stated in Prefaces, *A Defence of Poetry*, and elsewhere. In trying, then, to discover the ways in which the poems are 'trying to be good', all the chapters engage in detailed investigation of the works which are their subject. An allied purpose in doing so is to convey as adequate a sense as I can of the drama of Shelley's struggle with his language, his emotions, and his imagination. Throughout, comparisons with other poets—especially Wordsworth, Coleridge, and Byron—are used to illuminate Shelley's 'quality and importance'.

In opposition to accounts of Shelley that seek to ground his greatness as a poet in the coherence of his philosophical or political vision, I emphasize the frequency with which his best poems embody unresolved conflict, or dramatize the collision of perspectives, or engage in imaginings which chasten the reader's desire for cut-and-dried understanding. Earl Wasserman's study of the poetry, *Shelley: A Critical Reading*, is the chief monument of an interpretative tradition to which this book is indebted but with which it is frequently at odds. Specific disagreements with Wasserman's influential readings of the poetry occur throughout the book (as do disagreements—and agreements—with other critics). These disagreements are not intended to be point-scoring, but to clarify what is sacrificed as well as gained by a criticism that identifies poetic achievement with thematic coherence. Wasserman indicates the self-fulfilling nature of such criticism when he writes in his preface of 'the participation of the individual poem in the inclusive organizational principles of [the author's] "mind"'.[10] These 'inclusive organizational principles' and this 'mind' are critical hypotheses that, arguably, pay off handsomely: Wasserman's account of the poetry repeatedly demonstrates the fineness with which it is organized. But the weakness of *Shelley: A Critical Reading*, as Ross Woodman pointed out in a review, 'resides in the inevitable restriction imposed by what is ultimately a closed, self-confirming circle'.[11]

[10] Wasserman, p. viii. [11] *KSJ* xxi–xxii (1972–3), 245.

For Wasserman, Shelley's 'mind' is creative of and deducible from
the governing system of philosophical thought he finds embodied
in the poetry; for me, Shelley's 'mind' is inherent in and cannot be
abstracted from the frequently (and fruitfully) self-divided activity
of his language. By and large, Wasserman's book ignores tones and
emotional effects, the affective and imaginative workings of Shelley's
language; it discounts the presence of gaps and instabilities of
significance (it is revealing that his study does not include a reading
of *The Triumph of Life*, a poem which poses the most serious
challenge to the notion of an all-unifying metaphysic pervading the
poetry). Even when, as in the discussion of *Alastor*, the poem's 'art'
is defined as its 'protean evasion of any effort to pin it down into
an assertion', the expectation which 'protean' raises that the critic
will address the imaginative impact of the poem is swiftly quenched;
this 'protean' art serves a purpose which is strictly 'functional'.[12]
Wasserman does allow for the interplay of perspectives; but this
interplay is too often deprived of its capacity to move, affect, disturb,
or excite by the critic's discovery of some overall quasi-metaphysical
design to which it is subordinate. Most crucially, Wasserman seems
uninterested in the poetry's varying degrees of success.

That said, my emphasis on what in the previous paragraph I called
'instabilities of significance'—in, say, the discussion of the closing
lines of *Epipsychidion* or the account of the 'shape all light' in
The Triumph of Life—should not mislead the reader; this book has
only imperfect sympathy with deconstructionist readings of the
poetry. The poet may be 'a hero to adepts of Deconstruction';[13]
indeterminacy, the oscillation between possible meanings, is no
stranger to a poetic vision that wavers between scepticism and
idealism, and is acutely conscious of the gulf between language and
reality ('What thou art we know not; | What is most like thee?', 'To
a Sky-Lark', 31–2). Yet *The Triumph of Life* is a major poem not
because it lays bare the processes of figuration, but because its
presentation of complicated experience at once renders and respects
complication. As is often the case in Shelley's finest work, the poet's
imaginings make a passionately open-minded yet accomplished
response to the challenge of 'Life', an abstraction used with precise
awareness of the difficulty of defining experience. Shelley may be
the ostensible focus of *Deconstruction and Criticism*; but for all the

[12] Quotations from Wasserman, 38.   [13] 'Keach and Shelley', 12.

subtlety of the essays (especially the windings and unwindings of Paul de Man's piece on *The Triumph of Life*), their real concern is a more abstract obsession with textuality. The allegorizing crudeness this can lead to is evident in de Man's summary of his reading of the 'shape all light': 'We now understand the shape to be the figure for the figurality of all signification.'[14]

It is this tendency to slight the particularity of poems, to see them as exemplifying certain theories about language, which often flaws deconstructionist readings of Shelley and makes them, despite their alertness to contradictions, critically unpersuasive. As William Keach puts it, 'post-structuralist Shelleyans—stimulating as much of their work is—have been too little concerned with distinguishing the elusive activity peculiar to Shelley's writing from the problematic condition of language generally'.[15] Perhaps Jerrold E. Hogle's articles constitute the most exciting criticism of Shelley from a post-structuralist perspective.[16] But his reading of *The Witch of Atlas* offers what in my view is a representatively misleading and uncritical account of the poem as being concerned with 'the self-controlling potentiality of metaphor and metamorphosis'.[17] 'Misleading' because, as I argue in my chapter on the poem, *The Witch of Atlas* is, to a remarkable degree, highly self-aware about its self-awareness, using a supple array of devices to avoid being pinned down as simply concerned with its own processes; 'uncritical' because the essay appears to assume that reflexiveness is an unquestionable good.[18] A central aim of this study is to argue that Shelley's poems are keenly conscious of their own workings, but that this self-consciousness swings between linguistic self-absorption—sometimes fascinating, sometimes not—and engagement with what we are made to feel lies

[14] 'Shelley Disfigured', *DC*, 62.   [15] Keach, p. xii.
[16] See 'Shelley's Fiction: The "Stream of Fate" ', *KSJ* xxx (1981), 78–99, 'Shelley's Poetics: The Power as Metaphor', *KSJ* xxxi (1982), 159–97, and 'Metaphor and Metamorphosis in Shelley's "The Witch of Atlas" ', *SIR* xix (1980), 327–53. See also the fine study by Ronald Tetreault, *The Poetry of Life: Shelley and Literary Form* (Toronto, Buffalo, and London, 1987), which came to my notice after this book was written. Tetreault describes the poet as inviting 'the reader to participate in the endless pursuit of meaning through the play of *différance*', p. 16.
[17] 'Metaphor and Metamorphosis in Shelley's "The Witch of Atlas" ', 332.
[18] Throughout I use 'reflexiveness' or 'reflexivity' to mean a concern displayed by a poem with its own imaginative workings. By 'reflexive imagery' I mean, in William Keach's words, figurative constructions 'in which an object or action is compared, implicitly or explicitly, to an aspect of itself, or is said to act upon or under the conditions of an aspect of itself', Keach, 79.

beyond language even as it is being incorporated within language. The evaluative implications of this view are teased out throughout the book: they underpin the ways in which I interpret the fact that Shelley's language is often more intent on self-sustaining invention than on allegory; they emerge from the discussion of *The Cenci*'s self-consciousness about the relationship between language and psychological states—this self-consciousness is seen as riskily explicit but none the less impressive; they are apparent in the discriminations I make between the finest poetry of *Epipsychidion* and passages in that poem which border on verbal self-indulgence; and they are evident in my preference for those moments in *Prometheus Unbound* where the desire to create an imaginatively autonomous language blends with recognition of realities beyond the linguistic. Such passages most effectively answer Leavis's charge that Shelley's poetry is characterized by a 'weak grasp upon the actual'.[19]

Nevertheless, the issue of Shelley's self-awareness is complicated. The major limitation of William Keach's subtle *Shelley's Style* is that he presents a poet whose effects are always highly deliberated and therefore, it is at times assumed, successful. Sometimes one may agree with Keach that Shelley's use of language displays such deliberateness. An example is this famous passage from Act II of *Prometheus Unbound*:

> Hark! the rushing snow!
> The sun-awakened avalanche! whose mass,
> Thrice sifted by the storm, had gathered there
> Flake after flake, in Heaven-defying minds
> As thought by thought is piled, till some great truth
> Is loosened, and the nations echo round
> Shaken to their roots: as do the mountains now.
>
> (II. iii. 36–42)

Keach's analysis brings out the care with which Shelley allows his words to be read as conveying 'simultaneous natural and mental actions':[20] the avalanche and the revolutionary mental processes are not simply compared; they are made to intermingle. But just how much should we admire these lines? Valerie Pitt speaks of 'Shelley's major fault' as being that of 'over-elaboration',[21] and, for all the

---

[19] Leavis, 194.   [20] Keach, 77.
[21] One of a number of contributors to 'Reading Shelley', 'The Critical Forum', *EIC* iv (1954), 102.

cunning of the passage, there is a doggedness about the way it is worked out (especially apparent in the final line) that deprives it of the capacity to surprise. This is not to deny that the passage is full of analysable ingenuities, but to assert that the writing's design on us obtrudes too palpably. Though the meshing of flake-gathering avalanche and Heaven-defying minds is carried out thoroughly and precisely, it is willed: as a result, the key phrase, 'some great truth', seems little more than a listless abstraction. The best passages in *Prometheus Unbound* are less predictably determined by the writer's will. I have quoted Keach on the shortcomings of post-structuralist criticism of Shelley, yet Hogle, in turn, scores a hit when he rebukes Keach's book for its 'failure to see Shelley's style as a *giving way* to language while it is also a manipulation of it'.[22]

If at times, then, there is a will-driven feel about Shelley's language, there are also many moments when an escape from deliberateness (that is not merely a surrender to what the deconstructionist sees as the innate unreliability of language) manifests itself. A question posed in *Shelley's Style* is relevant to the restless and intriguing struggle between overall purpose and local texture occurring in many of the poems and explored throughout this study: 'Can Shelley's powerful sense of the poem as text or artifact ever fully accommodate his even more powerful sense of the poem as experience?'[23] In analysing 'Hymn to Intellectual Beauty' and 'Mont Blanc', for example, I discuss the rewarding presence of shifts of tone, jumps of and gaps between thought and feeling, and argue that 'self-awareness' is shown in the way the poems refuse to signpost or underscore one dominant meaning. This discussion involves consideration of the success with which the poems communicate difficult, mercurial states through language. Again, the chapters on *Alastor* and *Julian and Maddalo* distinguish between the poetry's readiness to dramatize contradictions and mere fuzziness or vagueness, faults which, for the most part, are absent from the poems. Once more, 'self-awareness' has to do with the way the reader is alerted to the writer's consciousness at work in the language, refusing to codify fluidities and suggestions. Lucy Newlyn presents us with what my book regards as too clean-cut a choice when she writes of *Alastor*: 'Either one sees the poet himself as sceptically distanced from the narrator, which gives him an immunity to solipsism (and the omniscient status,

---

[22] Review of *Shelley's Style*, *KSJ* xxxv (1986), 187.   [23] Keach, 119.

therefore, of a god). Or one sees the poet, like the narrator, as implicated in the reflexivity he ironises, which means that the poem self-deconstructs.'[24] It is Shelley's capacity to be both 'distanced' from and 'implicated' in the poem's experience which makes it unsettling, subversive of comfortable stances, imaginatively compelling. There are flaws in *Alastor* as my chapter acknowledges, and the poem is in places incompletely achieved. Yet those flaws derive more from the poem's occasionally intrusive attempts to supply a gloss on its imaginings than from its refusal to satisfy the critical hunger for coherence.

The presence in Shelley's language of creative forces at odds with the mere statement or amplification of predetermined themes prompts me to read the poetry as the reverse of didactic or propagandist. There are occasions when ideas are simply stated in the poetry; often these occasions represent the least rewarding moments in Shelley. More usually, ideas undergo imaginative transformation: poetry, writes Shelley in *A Defence of Poetry*, 'transmutes all that it touches, and every form moving within the radiance of its presence is changed by wondrous sympathy to an incarnation of the spirit which it breathes' (*PP* 505). I explore the achievement and cost of such transmutations, especially in the chapters on *Prometheus Unbound*, *The Witch of Atlas*, and *Epipsychidion*. My concern is not to deny the poetry's historical and political obsessions, but, by offering an analysis of the way its language works, to claim that its value has to do less with the unambiguous assertion of beliefs than with the imaginative testing and experiencing of ideas and emotions, ways of knowing and feeling. 'Tension', that 'much-used term relying on its context for whatever particular meaning it may have',[25] is a word I have found unavoidable in trying to pinpoint the behaviour of Shelley's language when it is most under stress and most interesting. Yet it is not the word's new critical suggestions of some harmonized equipoise of opposites that are valuable. Shelley's poems are characterized more by the sense that coherence and resolution are fictions, however necessary. This sense is most finely evident when accompanied, as is the case in different ways with *The Witch of Atlas*, *Epipsychidion*, and *The Triumph of Life*, by an admirably controlled

---

[24] 'Shelley's Ambivalence', review of *Shelley's Style*, *EIC* xxxvi (1986), 265.
[25] From the entry 'Tension' by Natan Zach in Roger Fowler (ed.), *A Dictionary of Modern Critical Terms* (1973; London, Henley, and Boston, 1982), 192.

inventiveness. Certainly the co-presence of conflicting forces makes for 'tension', but, as my readings seek to demonstrate, it is not tension of an equable, easily mastered kind.

In *The Witch of Atlas* the impulse to fictionalize and celebrate fictions is always threatening to unmoor itself from human interest; the poem's sense of and response to this threat contribute to its ability to surprise and affect. In *Epipsychidion* the desire to erase separateness (and embody that erasure in the language) accompanies and ultimately wars with the wish to affirm the uniqueness of another person. In *The Triumph of Life* the longing for certainty battles at crucial moments with the recognition of unknowableness. Along with the daringly ambitious if variable *Prometheus Unbound* and the subtly inconclusive *Julian and Maddalo*, these three poems from the poet's last years seem to me his most significant achievements; the book is organized more or less chronologically to point up Shelley's developing capacity to make great poetry out of conflicts.[26] Unavoidably, given my commitment to close readings of the poems I concentrate on, several excellent pieces such as 'Ode to the West Wind', the poems to Jane Williams, and the uneven but distinguished *Adonais* are touched on lightly or left out. However, though the book does not treat the whole of Shelley, its main purpose is to suggest ways in which a remarkable poet might be both fairly assessed and justly admired.

---

[26] For details of chronology see 'Outline of Shelley's Life', *PP*, pp. xvii–xix, as well as the notes about individual poems in the same edn. As will become evident, the ordering of Chapters Three, Four, and Five—on *Julian and Maddalo*, *The Cenci*, and *Prometheus Unbound*, respectively—is based on thematic as well as chronological considerations. The composition of these three works is closely related in time. (See *PP* for details.) For the view that at least some lines of *Julian and Maddalo* may have been drafted in 1818 (even though most of the poem was written in 1819) see G. M. Matthews, ' "Julian and Maddalo": The Draft and the Meaning', *Studia Neophilologica*, xxxv (1963), 57–84, esp. 65–6.

1

Alastor
The Fluidities of Narrative

Milton Wilson pinpoints the principal evaluative problem which Shelley's first long poem of distinction raises when he asserts: '*Alastor* is a profoundly ambiguous poem, not because it is a good and complex poem, but because Shelley has failed to work it out with sufficient rigor and finality.'[1] However, the poem's reluctance to use poetry as a problem-solving vehicle is evident, while its suspicion of 'finality' is central to its achievement. *Alastor* thrives imaginatively on contradiction and evasion, displaying an almost Calvino-like awareness that reality rarely conforms to the spirit's maps and charts. Yet the Romantic poet allows us to experience on our pulses an insight which, in *Mr Palomar*, the sophisticated modernist novelist spells out: 'But if for a moment he stopped gazing at the harmonious geometrical design drawn in the heaven of ideal models, a human landscape leaped to his eye where monstrosities and disasters had not vanished at all and the lines of the design seemed distorted and twisted.'[2]

In a comparable way the 'Poet' of Shelley's narrative 'seeks in vain for a prototype of his conception' (*PP* 69). Two points immediately rise. First, reference to the 'Poet', a title which Shelley employs in the Preface and the poem, prompts us to ask whether *Alastor* presents us with different 'characters', an issue that is both fascinating and vexed. In my view, Earl Wasserman's account of the poem—he sees it as dramatizing a split between a Wordsworthian 'Narrator' who believes in communion with nature and a Shelleyan 'Poet' (or 'Visionary') who is driven by the 'desire for a transcendent good'[3]—supplies too clear-cut a dialectical design, one which ignores the degree to which the two figures overlap, blend, are involved in one another. And, secondly, quite what meaning we

---

[1] *Shelley's Later Poetry: A Study of his Prophetic Imagination* (New York, 1959), 164.
[2] Trans. William Weaver (London, 1986), 98.   [3] Wasserman, 41.

should attach to 'prototype' is unclear. Evan K. Gibson interprets it thus: 'Shelley does not say that the poet seeks for a copy of the vision in the actual world. . . . he seeks for the pattern or original of the vision itself.'[4] Yet Shelley—in both Preface and poem—does not make the kind of sharp distinction between 'real' and 'ideal' on which Gibson depends, a refusal to distinguish that makes the poem the more intriguing, explains its hold over the reader. *Alastor* involves itself in murkiness, unsureness, exploration—and often benefits poetically from doing so.

Should we find fault with the Poet for his 'self-centred seclusion' (*PP* 69)? Or should we admire the idealism he displays in pursuing the 'vision in which he embodies his own imaginations' (*PP* 69)? Though the Preface elicits both responses from the reader, it privileges neither; it launches a poem which is neither sentimental nor censorious in its overall effect. Shelley offers a poetry compelled by its deepest compulsions (which are not deconstruction's all-governing laws) to 'keep its transformations going in unconfined ripples of interpretation within the poem and after it'.[5] C. E. Pulos prefers a more didactic coherence:

Shelley implies in *Alastor* that Beauty, so far as we know, has no objective existence. The tragedy of the hero of *Alastor* lies in his failure to realize this conviction of his creator. Instead of looking for the likeness of his vision in a human maiden, the hero of *Alastor* vainly seeks to apprehend its pattern in ultimate reality.[6]

Pulos argues the poem into the philosophical clear-sightedness he values by berating the 'hero' for failing to grasp a 'conviction of his creator'. But this 'conviction' is of Pulos's making, not Shelley's. Shelley describes the tragic career of his protagonist in a more guarded way: rather than moralizing—'the hero . . . vainly seeks'—he sounds a note which is sorrowful, elegiac: 'He seeks *in vain* [my italics] for a prototype of his conception.' (*PP* 69.)

---

    [4] '*Alastor*: A Reinterpretation', *PP* 549, reprinted from *PMLA* lxii (1947), 1022–45.
    [5] Jerrold E. Hogle, 'Metaphor and Metamorphosis in Shelley's "The Witch of Atlas"', 329.
    [6] *The Deep Truth: A Study of Shelley's Scepticism* (1954; Lincoln, Nebr., 1962), 81.

Seeming to turn Wordsworth's opposition in *The Excursion* to 'wilful disesteem of life | And proud insensibility to hope'[7] against the older writer's generation, Shelley does engage in polemic in the Preface. But the polemic shows signs of centrifugal strain; it hits out at a class, those 'deluded by no generous error, instigated by no sacred thirst of doubtful knowledge, duped by no illustrious superstition, loving nothing on this earth, and cherishing no hopes beyond' (*PP* 69), that makes no appearance in the poem. Shelley is not so much trying to sum up the meaning of the poem he had written as attacking attitudes to which the experience explored by the poem clarifies his opposition. But his language of condemnation—'instigated by no sacred thirst of *doubtful* knowledge', '*duped* by no illustrious superstition' (my italics)—is fully alive to the pitfalls involved in the implicitly preferred alternative.

However, conflict of feelings such as the Preface displays is the source of the poem's power. The Preface elegizes rather than admonishes, but elegizes tough-mindedly. Shelley is alive to the almost unavoidably self-destructive courses through which the 'pure and tender-hearted' are driven 'when the vacancy of their spirit suddenly makes itself felt' (*PP* 70). The first sentence of the Preface, cool and curious, advertises the inquiring nature of what is to follow: 'The poem entitled "ALASTOR," may be considered as allegorical of one of the most interesting situations of the human mind.' (*PP* 69.) The poem is 'allegorical' not because its details can be translated easily into abstractions, but because it is engaged in an attempt to find verbal equivalents for inner states. And when Shelley speaks of 'one of the most interesting situations of the human mind', he strikes the tone which governs his poem: that of penetrating, unmoralistic exploration. The reader is led, as in other major poems by Shelley (*Julian and Maddalo* and *The Triumph of Life* are examples), through a labyrinth which leads, simultaneously, towards a cul-de-sac and an abyss: a cul-de-sac in that no way of escape from the poem's dilemmas presents itself; an abyss in that dizzying possibilities of interpretation offer themselves to us. *Alastor*

---

[7] iv. 1029–30. See Paul Mueschke and Earl L. Griggs, 'Wordsworth as the Prototype of the Poet in Shelley's *Alastor*', *PMLA* xlix (1934), 229–45: they present an over-simplified view of the poem as 'allegorical of one poet's [Shelley's] interpretation of a contemporary poet's [Wordsworth's] fate', 245. But Shelley undoubtedly did see *The Excursion* as showing, in the words of Mueschke and Griggs, 'Wordsworth's defection from the cause of liberty', 234.

leaves its reader in possession of an experience which is at once claustrophobic and vertiginous.

The invocation (1–49) is an impressive piece of blank verse which feels as though it is being voiced by Shelley, for all the commentators' discovery of a mixed-up Wordsworthian Narrator.[8] As has often been observed, Wordsworth has influenced this passage; cadence and diction bear witness to Shelley's immersion in the older poet's work, especially *The Excursion* and 'Ode: Intimations of Immortality' ('Immortality Ode').[9] But it is a fine passage which attains its own kind of authority. It is not merely under Wordsworth's shadow; it is able to cast a shadow over Wordsworth. Nature as 'Mother of this unfathomable world' (18) is a far less benign presence in Shelley than in Wordsworth. Where Wordsworthian Nature offers security and comfort, stays against uncertainty, Shelleyan Nature arouses fear and awe, emotions associated with the 'sublime', with awareness of the mind's loneliness in an unknowable universe. Loneliness is the mood which dominates these lines, despite their plea that the self should be subsumed within the larger community of nature, the 'beloved brotherhood' (1) which the speaker invokes wish-fulfillingly. The concluding lines illustrate the skill with which Shelley infuses Wordsworthian sentiment with something more homeless:

>   and, though ne'er yet
> Thou hast unveil'd thy inmost sanctuary,
> Enough from incommunicable dream,
> And twilight phantasms, and deep noonday thought,
> Has shone within me, that serenely now
> And moveless, as a long-forgotten lyre
> Suspended in the solitary dome
> Of some mysterious and deserted fane,
> I wait thy breath, Great Parent, that my strain
> May modulate with murmurs of the air,
> And motions of the forests and the sea,
> And voice of living beings, and woven hymns
> Of night and day, and the deep heart of man.
>
>   (37–49)

Echoes of 'Tintern Abbey'—its famous climax (93–102)—cannot offset the fact that Shelley's pantheism is more gesture than substance,

---

[8] Principally Wasserman in his opening ch.
[9] See Mueschke and Griggs, esp. 235–6 and 238.

offering the thinnest of veils for a troubled subjectivity. If in 'Tintern Abbey' Wordsworth claims to have felt 'A motion and a spirit' (100), and by virtue of his modulated blank verse is able to persuade us that this claim is authentic, Shelley is far less assertive. It is his achievement to make us hear pathos just behind affirmation. The passage opens with a confession of failure—Nature's 'inmost sanctuary' still eludes the speaker's desperate researches; it proceeds to put its weight on a very untrustworthy adjective, 'incommunicable dream'; its first reference to inspiration is unabashedly internalized, 'Has shone within me'; and its second uses a figure of speech (42–4) whose suggestions—of 'solitariness' and experience of neglect—work quite unserenely on the reader's imagination. There is, then, throughout the passage an undertow of darker feelings at odds with the speaker's insistence that he awaits the inspiring breath of Nature in a state which is 'serene'.

The invocation half-expects no response; it is an incantation made by a poet who wishes to escape from solipsism into communion with external reality, but who fears or knows that such an escape is improbable. He then launches abruptly into a tale which bears out the truth of this darker recognition, one which the reader hears almost between the lines of the invocation. The speaker would like to rehouse Nature within his imagination, to make it homely. However, there is an eerie gap between the poetry's professions of warmth and love towards Nature and the silence which surrounds these professions. Freud sees the 'uncanny [*unheimlich*]' as 'something which is secretly familiar [*heimlich-heimisch*], which has undergone repression and then returned from it'.[10] In *Alastor* it is the fear that Nature does betray the heart that loves her which has undergone repression and then returned during and beyond the invocation. The speaker evokes a vision of community; it merely intensifies his isolation. He wishes to enter into ghostly dialogue with the natural, but hears only his own monologue, or such speech as is the projection of his own imagination. In fact it is the hearing of silence which is most imaginatively gripping: 'In lone and silent hours, | When night makes a weird sound of its own stillness' (29–30).

---

[10] 'The "Uncanny"', *'An Infantile Neurosis' and Other Works* (1955), *The Standard Edition of the Complete Psychological Works of Sigmund Freud*, ed. James Strachey, trans. in collaboration with Anna Freud, assisted by Alix Strachey and Alan Tyson (24 vols., London, 1953–74), xvii. 245.

In the preceding lines a Wordsworthian echo—the reference to
'obstinate questionings' the speaker hopes to 'still' (26)[11]—serves to
define the distance between the two poets:

> I have made my bed
> In charnels and on coffins, where black death
> Keeps record of the trophies won from thee,
> Hoping to still these obstinate questionings
> Of thee and thine, by forcing some lone ghost,
> Thy messenger, to render up the tale
> Of what we are.
>
>   (23–9)

Wordsworth gives thanks for 'obstinate questionings' since they
confirm his intuition of a prior existence; they prompt trust in the
Platonic myth of pre-existence by means of which he is able to stave
off his fear of imaginative failure. Shelley's 'obstinate questionings'
are not in themselves earnests of good so much as admissions of
incomplete knowledge. His lines fuse opposing impulses: a will-
driven desire to know and a sense that no answer has yet been (or
ever will be) forthcoming; the 'tale | Of what we are' remains a 'tale',
a fiction to be ravelled and unravelled in the larger tale which is
Alastor. What begins as melodrama ('charnels and . . . coffins') ends
as something stranger, a brooding over the fictionalizing involved
in knowing. These changes of register should not be seen as the
blemishes of a poet fumbling with a new style. Rather, they underline
shifts in feeling. *Alastor* is at once an unstable poem and a poem
in control of its instabilities. But this control does not take the form
of a demarcated gap between Shelley and his 'characters'. In what
follows reference shall, for the sake of convenience, be made to
the Narrator and the Poet, but ultimately both figures are barely
separable from Shelley's self-examining imagination. In *Alastor* that
imagination projects figures, narrative situations, and images in an
attempt to body forth its obscurest impulses and to see them as other.

The poem's fluidity stems from Shelley's reluctance to write
straightforwardly autobiographical poetry or straightforwardly
objective narrative. Self-exploration is given a narrative form;
narrative is made the vehicle of self-exploration. Throughout his
career Shelley uses narrative in highly individual ways; he eschews

---

[11] Cf. 'Immortality Ode', 145.

story and favours reflexiveness. Indeed, only the difficulty at crucial moments in the poem of distinguishing between 'Shelley' and a separate 'Narrator' prevents assent to William Keach's persuasive view of the poem and its use of narrative: 'The wandering poet may be seen as the narrator's deeper self projected as spectral other . . . When we read the poem this way, the entire elegiac narrative becomes a symbolic fiction in which the wandering poet, the dream vision and the landscape function as projections of the narrator's troubled psyche.'[12]

Keach's argument attempts to retrieve order out of a text of great elusiveness, to find a centre for the poem's fictions in the all-controlling consciousness of the 'Narrator'. But it involves the critic himself in the creation of a fiction, this 'all-controlling consciousness'. The poem will not fold back obligingly into such an origin of significance. The writing insists on how much is withheld from the Narrator's understanding. Despite parallels between the Narrator's researches and the Poet's, there is, in the account of the latter, a distinct privateness, a self-enclosed quality: the Poet seems lost in enchantment as he 'ever gazed | And gazed, till meaning on his vacant mind | Flashed like strong inspiration, and he saw | The thrilling secrets of the birth of time' (125–8). Here the word 'vacant' does not evoke suggestions of imaginative death as it does in a later use at line 195, where the Poet wakes from his vision to behold 'The distinct valley and the vacant woods'. Rather, it implies a mind emptied of anxiety, contemplative, beyond the reach of the Narrator's imaginings. Again, the dream-vision is less a projection from the Narrator's troubled psyche than a process in which reflexiveness and otherness jostle for the upper hand:

> A vision on his sleep
> There came, a dream of hopes that never yet
> Had flushed his cheek. He dreamed a veiled maid
> Sate near him, talking in low solemn tones.
> Her voice was like the voice of his own soul
> Heard in the calm of thought; its music long,
> Like woven sounds of streams and breezes, held
> His inmost sense suspended in its web
> Of many-coloured woof and shifting hues.
> Knowledge and truth and virtue were her theme,

---

[12] Keach, 87.

> And lofty hopes of divine liberty,
> Thoughts the most dear to him, and poesy,
> Herself a poet.
>
> (149–61)

The texture of the narrative partakes of the poem's larger ambivalences. So this passage begins and ends with the objective tone appropriate to an account of something which 'really happened'. And yet this objective tone is enmeshed in suggestions which imply the subjective nature of the experience. Shelley cunningly phrases the opening lines to draw attention to the fact of the 'vision' rather than to the presence of a 'visionary' who dreamed it up. But within a few lines we are taken into a dream-encounter with a 'veiled maid' who is markedly more phantasmal and inseparable from the Poet's imaginings than later Shelleyan anima-figures such as Emilia in *Epipsychidion*.

In *Alastor* Shelley's blank verse pauses thoughtfully over ambiguities: 'Her voice was like the voice of his own soul | Heard in the calm of thought'. Does that 'like' separate or unite the two voices? The success of the writing derives from the fact that Shelley has found an idiom which is attuned to his own uncertainties: 'projected self-reflection'[13] the veiled maid may be, but the language locates us inside a state of 'reverie' such as Shelley would describe in *On Life* (1819): 'Those who are subject to the state called reverie feel as if their nature were dissolved into the surrounding universe, or as if the surrounding universe were absorbed into their being.' (*PP* 477.) That the passage concludes with the phrase 'Herself a poet' sustains the unsureness Shelley has built up in the reader's mind: the phrase concedes an independent identity to the veiled maid even as it gives fresh support to the view that the Poet is creating an ideal image (the 'single image' of the Preface, *PP* 69) with which to fall in love.

The pathos achieved by the ensuing lines comes, in fact, from Shelley's creation of an experience in which distinctions between what is 'real' and what is 'imagined' no longer obtain. Fantasy and actuality meet, blend, and blur:

> at the sound he turned,
> And saw by the warm light of their own life
> Her glowing limbs beneath the sinuous veil

[13] Keach, 82.

Of woven wind, her outspread arms now bare,
Her dark locks floating in the breath of night,
Her beamy bending eyes, her parted lips
Outstretched, and pale, and quivering eagerly.
His strong heart sunk and sickened with excess
Of love.

(174–82)

What is imagined here will not be dismissed as (merely) 'autoerotic'
or (simply) 'projection'. Keach claims in relation to the much-
discussed second line that the 'reflexive locution signals a self-inclosed
psychical experience'.[14] This certainly defines part of the poetry's
impact. Yet there is a danger of creating sharp divisions where Shelley
chooses to intrigue and tantalize. In fact, the phrase 'by the warm
light of their own life' does not so much refer the light back to the
mind of the perceiver as insist on the 'otherness' of the maiden's
'glowing limbs'. This 'otherness' has to compete with suggestions
that do support the idea of a 'self-inclosed psychical experience':
the disembodying, across the line-break, of the 'sinuous veil' which
turns out to be made of 'woven wind'; the fluidity of dream created
by participles, 'dissolving' rhythms; the similarity between the
description of the maiden 'quivering eagerly' and the Poet overpowered
by 'excess | Of love'. Shelley forfeits the kind of analytical objectivity
his critics favour; in its place he supplies a poetic experience which
is precisely rendered in its own terms and allows us, non-logically
perhaps, to entertain two conflicting notions at the same time: that
the veiled maid both is and is not a being separate from the Poet.
Such non-logical fusions of opposing feeling are found quite often
in Shelley's greatest poetry, culminating in the evocation of the 'shape
all light' in *The Triumph of Life*. That said, the passage just looked
at anticipates rather than participates in greatness. It is too adjectival,
a stylistic flaw which suggests an insistence more apparent in the
ensuing lines (182–91), where one is torn between thinking that
Shelley is overwriting and that he is staying close to a feverish
intensity.

This encounter between the Poet and the 'veiled maid' illustrates
how the success of *Alastor*'s best passages depends on Shelley's
capacity to involve us in the poem's experience rather than on

[14] Keach, 82.

certainty about how this experience should be interpreted. Is such
involvement aroused in the reader by the ensuing sections of the
poem? Undoubtedly Shelley handles the aftermath of the vision
skilfully. There is an effective change of tempo: from the complex,
inwoven rhythms of the preceding lines Shelley shifts to a starker
style:

> Roused by the shock he started from his trance—
> The cold white light of morning, the blue moon
> Low in the west, the clear and garish hills,
> The distinct valley and the vacant woods,
> Spread round him where he stood. Whither have fled
> The hues of heaven that canopied his bower
> Of yesternight?
>
> (192–8)

Distinctness, clarity: these 'Immortality Ode'-like symptoms of
loss are evoked with admirable mimetic terseness in the opening
description. The Wordsworthian debt is acknowledged openly in the
question which concludes these lines.[15] If Shelley's mood is very
much his own, one way of marking his independence from the older
poet is to observe that he is composing a poetry to which it is not
accurate to apply a phrase such as 'Shelley's mood'. Much that is
fascinating about *Alastor* derives from the way it intermingles
narrative and lyrical effects. Narrative is not always a helpful
presence; the detachment it makes possible gives rise to attempts to
close the emotional gap between Narrator and Poet. Such an attempt
leads in the passage from lines 207–19 to writing which, though
powerful and even haunting in its questioning of unanswerable
absolutes, risks overstatement. More seriously, the lure of authority
which narrative holds out to an author—that he can insert pointers
towards meaning—may be seen at work in these lines: 'The spirit
of sweet human love has sent | A vision to the sleep of him who
spurned | Her choicest gifts' (203–5). Despite the episode of the Arab
maiden, there has been little in the poem to warrant this abrupt
gesture towards offering a moral. The lines are best seen as Shelley's
attempt to impose a meaning on a poem whose complexity will not
be reduced to a tale of retribution for the crime of 'self-centred
seclusion'.

---

[15] Cf. 'Immortality Ode', 56–7.

Yet two advantages of Shelleyan narrative suggest themselves. One is its quickness to make jumps which may floor the theme-constructing critic, but ought to excite his imagination. For example, the speed with which the Poet moves from the post-vision disenchantment to a sense that death may hold all the answers is remarkable and yet easily overlooked because of the momentum of the verse. The effect of this jump is crucial. Now that 'death' is part of the poem's dramatis personae certain emotional possibilities (discussed more fully below) lie to Shelley's hand. A second advantage of Shelleyan narrative is the way it can give a sharp edge to what in a lyric poem would risk self-pity. The lines which evoke the Poet's state of 'vacancy'—this time the vacancy of loss and disillusion—are an example: 'His wan eyes | Gaze on the empty scene as vacantly | As ocean's moon looks on the moon in heaven' (200–2). The description is more able to affect by reason of its detachment, its air of thoughtful 'looking on'.

It would be wrong to contend that this level of achievement is sustained throughout the poem. Passages from the long sections (222–671) describing the Poet's forlorn wanderings, his supra-natural boat-journeys, and protracted dying may seem flawed in various ways. Yet often worry about some aspect of the writing coexists with recognition of counterbalancing strengths. The risk that the Poet's experience is overstated, in danger of losing its power over the reader through sheer insistence, is real, and there are moments when risk topples into fact. That the Poet's 'hope' (221) affected him 'like despair' (222) is a paradox worth offering once. When seventy lines later Shelley tells us of the Poet that 'A gloomy smile | Of desperate hope wrinkled his quivering lips' (290–1), his imagination is clearly treading water.

And it is true that the balance achieved in the writing at lines 200–2 can seem to falter in what follows. A passage such as this, for example, is calculated to prompt a 'yes (or no), but . . . ' from the reader:

> He wandered on
> Till vast Aornos seen from Petra's steep
> Hung o'er the low horizon like a cloud;
> Through Balk, and where the desolated tombs
> Of Parthian kings scatter to every wind
> Their wasting dust, wildly he wandered on,
> Day after day, a weary waste of hours,

> Bearing within his life the brooding care
> That ever fed on its decaying flame.
> And now his limbs were lean; his scattered hair
> Sered by the autumn of strange suffering
> Sung dirges in the wind; his listless hand
> Hung like dead bone within its withered skin;
> Life, and the lustre that consumed it, shone
> As in a furnace burning secretly
> From his dark eyes alone.
>
>   (239–54)

Shelley's narrative design commits him to an extended account of the Poet's wanderings. In the best sections of this part of the poem, he wrests subtle intimations from description and incident. Yet if the larger structural shape of the poem is persuasive, its details are less so. Certain effects look cheap, especially those contrived by over-use of words such as 'strange' (a point made by Edmund Blunden who remarks: 'An imperfect command of the diction is . . . betrayed in such details as the repeated application of an adjective . . . until particularity is lost').[16]

The melancholy vowel-music and repetitions may seem to be laid on too ostentatiously for the passage to affect the reader deeply. Yet the accents of suffering emerge from, almost overcome, crudenesses in the writing. Here, for instance, Shelley communicates a strong sense of the self being consumed by its deepest energies. He conveys this through the image which concludes the passage and through these lines, with their mixture of positive and negative suggestions: 'wildly he wandered on, | Day after day, a weary waste of hours, | Bearing within his life the brooding care | That ever fed on its decaying flame'. 'Brooding' draws to itself the ambivalences which surround the Poet; the word suggests creative incubation and moody pessimism. The passage illustrates both the poem's unevenness and its frequent impressiveness.

Alastor shares with other long poems by Shelley—*Prometheus Unbound, The Witch of Atlas,* and *The Triumph of Life* particularly—two competing impulses: to press home a central idea, mood, or emotional complex on the one hand, to lose itself in the creation of fictions which spin away from the poem's supposed centre on the other. This quality in Shelley's work is a symptom of intricacy,

---

[16] *Shelley: A Life Story* (1946; London, New York, and Toronto, 1965), 114.

inward tension, of an imagination mistrustful of the certainties by which it is propelled. The central section of *Alastor* hovers between elusive symbolic or allegorical glosses and description which is there for its own sake. Atmospherics shot through with brief but pointed references back to the poem's main concern are what lines 316 to 369 have to offer. The writing is representative not of the greatest poetry in *Alastor* but of the typical. Impressive pastiche of eighteenth-century poets like Gray and Collins declares itself in places ('Twilight, ascending slowly from the east . . . ' (337 and following)); in others (such as 340–4) a fascination with sub-Miltonic sound and fury, less happy because too declamatory, emerges. What removes the passage (and the poem) from the literary museum is the presence of lines which focus tersely on the consciousness at the centre of this symbol-ridden landscape:

> As if their genii were the ministers
> Appointed to conduct him to the light
> Of those beloved eyes, the Poet sate
> Holding the steady helm.
>
> (330–3)
>
>   The little boat
> Still fled before the storm. . . .
>   Safely fled—
> As if that frail and wasted human form
> Had been an elemental god.
>
> (344–5, 349–51)
>
>   'Vision and Love!'
> The Poet cried aloud, 'I have beheld
> The path of thy departure. Sleep and death
> Shall not divide us long!'
>
> (366–9)

What intrigues about the first excerpt is the admission made by 'as if' that this is just one, probably fanciful, way of interpreting the Poet's journey. The irony is that 'their genii' refer to the spirits which preside over the threatening waves and are as likely to 'conduct' him to the realm of death as to those 'beloved eyes'. 'Conduct' was the verb used twice (212 and 219) by Shelley in the earlier meditation whether Nature leads to death or whether death leads to sleep, and something of the previous irresolution carries over into these lines. The Poet is at once passive sufferer and intrepid spiritual adventurer.

That death may be the hiding place of the 'beloved eyes' is a recognition which is always kept subliminally in mind, though not spelled out—and then ambiguously—until the Poet's own death. Whether this involves Shelley in a glamorizing of death which seems immature is arguable. Certainly he has turned away from another kind of tragedy to which he may allude by quotation in his Preface (*PP* 70): that of the 'sore heart-wasting' of Margaret in *The Excursion* (i. 875). Yet the Poet's 'wasting' has power to affect the reader not because it is made up of understated human details which touch the heart in a Wordsworthian way, but because it involves imagining an extreme: a simultaneous surrender to and defiance of disappointment. The poem's element of immaturity comes through in the third excerpt where the Poet's cry makes nonsense of its four highly emotive nouns ('Vision and Love', 'Sleep and death')—nonsense because the lines have a blurb's simplifying crudity. Something, though, of *Alastor*'s power to move is manifest in the second excerpt. There the switch of perspectives—'frail and wasted human form' in one line, 'elemental god' in the next—is handled with panache. It is a moment when the poem's complications of viewpoint both come to a head and find resolution. The lines solve nothing, but they express memorably the twin sense of the Poet as victim and hero, and in doing so they offer a brief cathartic release.

There are two different kinds of poetry in this section: a poetry that offers what Lloyd Abbey calls 'a symbolic recapitulation of the entire action of the poem'[17]—a way of writing which tends to spell out significance and borders on allegory—and a poetry which does not recapitulate and reflect, but advances and explores. The two kinds coexist amicably, but there is little doubt as to which is better. They can both be seen at work in the passage between lines 469 and 514, which contains one of the finest moments in the poem, an account of a 'Spirit' that is and is not identifiable with a subsequent apparition possessing 'Two starry eyes':

> A Spirit seemed
> To stand beside him—clothed in no bright robes
> Of shadowy silver or enshrining light,
> Borrowed from aught the visible world affords
> Of grace, or majesty, or mystery;—

---

[17] *Destroyer and Preserver: Shelley's Poetic Skepticism* (Lincoln, Nebr., and London, 1979), 27.

But, undulating woods, and silent well,
And leaping rivulet, and evening gloom
Now deepening the dark shades, for speech assuming
Held commune with him, as if he and it
Were all that was,—only . . . when his regard
Was raised by intense pensiveness, . . . two eyes,
Two starry eyes, hung in the gloom of thought,
And seemed with their serene and azure smiles
To beckon him.

                                            (479–92)

This passage occurs after a well-managed change of key at line 469
('Hither the Poet came'); the Poet's disturbance of and exclusion from
natural tranquillity are evoked in these lines:

                        His eyes beheld
Their own wan light through the reflected lines
Of his thin hair, distinct in the dark depth
Of that still fountain . . .

                                            (469–72)

The collapsing of subject and object across the line-ending—'His eyes
beheld | Their own wan light'—suggests the Poet's absorption in his
vision (what we are tempted to moralize as his 'narcissism', though
Shelley avoids such a judgement). This absorption modulates subtly
into the awareness of some other presence which the ensuing lines
(quoted above) manifest. Indeed, the fluidity in places of Shelley's
syntax can be defended as a means of suggesting the Poet's state of
reverie, a state in which hard and fast distinctions between 'Spirit'
and veiled maid (such as Carlos Baker insists on) are scarcely
applicable. Baker asserts that Shelley 'intends . . . to represent two
visions in conflict'.[18] And certainly lines 487–8 ('as if he and it |
Were all that was,—only . . . ') form a bridge between an intuition
of one kind of being and another. Yet this bridging serves rather
to connect the visions than to separate them. Shelley suggests
such a connection in his two uses of the word 'seemed' (479 and
491), which is applied both to the Spirit and to the possessor of
the 'starry eyes'. Both apparitions could, probably do take their
origin from the same source: the 'intense pensiveness' of the Poet.
Both participate in one another's mode of being. The 'Spirit' may

                    [18] Baker, 58.

represent 'phenomenal nature',[19] but it is represented as a personification, a visionary presence; the 'starry eyes' may indeed represent a ghostly return of the veiled maid to the Poet, but the writing suggests artfully that they may be 'merely' stars, endowed with visionary life by the Poet's imagination. What is valuable is Shelley's readiness to stay close to the shifting instability of the Poet's intuitions; that the 'starry eyes', for example, 'hung in the gloom of thought' blurs normal distinctions between 'inner' and 'outer'. Yet Shelley also complicates, by allowing for a double reading of, the Poet's state of mind, referring in one sentence to 'the gloom of thought', in the next to 'the light | That shone within his soul' (492–3).

Such suggestions—here the twining round one another of 'gloom' and 'light' as metaphors for the Poet's inward state—are the more impressive for their underplayed quality. The same is not true of the more spelled-out poetry referred to above, passages like the following monologue of the Poet which make for an easier but more pedestrian reading experience:

> O stream!
> Whose source is inaccessibly profound,
> Whither do thy mysterious waters tend?
> Thou imagest my life. Thy darksome stillness,
> Thy dazzling waves, thy loud and hollow gulphs,
> Thy searchless fountain, and invisible course
> Have each their type in me: and the wide sky,
> And measureless ocean may declare as soon
> What oozy cavern or what wandering cloud
> Contains thy waters, as the universe
> Tell where these living thoughts reside, when stretched
> Upon thy flowers my bloodless limbs shall waste
> I' the passing wind!
>
> (502–14)

One way of describing what is wrong with these lines is to say that they are an interpreter's gift. Lloyd Abbey, for example, sees them as anticipating the complex metaphysics of 'Mont Blanc': 'While the poet of "Alastor" does not perceive the existential interdependence of mind and stream, he nevertheless approaches this awareness in his realization that the flow of the stream typifies the flow of his perceptions.'[20]

---

[19] Baker, 58.   [20] *Destroyer and Preserver*, 28.

Undoubtedly Shelley is grappling with ideas about the source and destination of existence which he will refine in his later prose and poetry. Yet here the wording given his ideas is unimpressive, relying on obvious epithets ('profound', 'mysterious', 'searchless', 'measureless'), on a language of 'type' and emblem. To escape clumsiness such a language needs to show that it knows it is a language, one which constructs the real in the image of its own desires. Elsewhere in the poem, as in the reference to 'the tale | Of what we are', Shelley displays this kind of awareness. But like the Poet's earlier address to the swan (280–90) these lines sacrifice implication to statement, suggestiveness to reductiveness. The Poet has more power to interest the reader when caught in half-profile than when he addresses his predicament in an operatic, head-on way.

More satisfying are moments which insinuate symbolic implications as in the account of the 'pine' (561) which 'stretched athwart the vacancy | Its swinging boughs' (562–3). There the image is independent of but bears on the Poet's predicament. And the hero's doubt about the ultimate destination of existence can benefit from being mediated indirectly as when the river is described 'Scattering its waters to the passing winds' (570). Again, the account of the Poet's death blends evocation with implication:

> his last sight
> Was the great moon, which o'er the western line
> Of the wide world her mighty horn suspended,
> With whose dun beams inwoven darkness seemed
> To mingle. Now upon the jagged hills
> It rests, and still as the divided frame
> Of the vast meteor sunk, the Poet's blood,
> That ever beat in mystic sympathy
> With nature's ebb and flow, grew feebler still:
> And when two lessening points of light alone
> Gleamed through the darkness, the alternate gasp
> Of his faint respiration scarce did stir
> The stagnate night:—till the minutest ray
> Was quenched, the pulse yet lingered in his heart.
> It paused—it fluttered.

(645–59)

Perhaps the fading away is sentimentally protracted. Yet there is art in the way Shelley marries the Poet's dying with the gradual extinction of the moon's 'two lessening points of light' which recall, but recall

tactfully, the two starry eyes of the Poet's earlier vision. The poetry
has interested itself in imagining the scene as richly as possible; the
delicate modulations of the rhythms bear witness to this, as does
the sustaining of doubt about the Poet's relationship with Nature
to the very end. Shelley gets it both ways: the Poet is at once driven
beyond what the natural world can offer and profoundly attuned
to its 'ebb and flow'. There is no need to posit a mistaken Narrator
here, projecting his Wordsworthian views upon the Poet. The writing
does not alert us to interpretative error in this way; rather, it calls
a temporary truce, an end to fretting over unanswerable questions
in the face of a death which is evoked in these lines as release and
in the ensuing lines as utter cessation ('the murky shades involved
| An image, silent, cold, and motionless' (660–1)). It is the movement
towards and revulsion from the consolations of elegy which we hear
in the Narrator's final address. He spurns yet in spurning calls upon
the assistance of 'Art and eloquence' (710) to help him cope with
grief and absence; he attempts to deny artifice (referring to 'the frail
pauses of this simple strain' (706)), yet in so attempting admits his
dependence on it. The writing's conflicts are registered by its twists
and turns, by its use of a Wordsworthian tag (compare 'Immortality
Ode', 207) that asserts the prior claims of the inexpressible over
expression: 'It is a woe "too deep for tears"' (713). 'The tale | Of
what we are' which *Alastor* has attempted to narrate has led only
to the knowledge that things 'are not as they were' (720).

   If the theme of the poem is failure, critical judgement of *Alastor*
should pronounce it a modified success. At the poem's heart is loss
and unavailingly compensatory quest. The loss of the veiled maiden
leads to a complete crisis in the Poet's relations—epistemological,
imaginative, and emotional—with reality. The discontent it arouses
in him is almost religious, a desire for the numinous. In a puzzling
sentence from the Preface, Shelley refers to 'that Power which strikes
the luminaries of the world with sudden darkness and extinction,
by awakening them to too exquisite a perception of its influences'
(*PP* 69). But any such 'God term' is markedly absent from the poem
(save for the *Queen Mab*-like scorn of lines 675–81); and to read
this sentence as anticipating Shelley's sceptical affirmations at the
end of 'Mont Blanc' ('the power is there' (127)) would be over-
confident. Half an abstraction that represents an inward human
impulse, half a gesturing towards some transcendent force, the
'Power' of the Preface takes its place along with other hoverings in

the poem. Such hoverings show that Shelley was not entirely sure
how to interpret his poem's experience; his readiness to borrow
Peacock's proposed title is another example. Though Peacock did
Shelley a good turn in one way, in another he did him no kindness
by imputing to his friend a moralistic rigour which is not borne out
by the poem: 'He was at a loss for a title, and I proposed that which
he adopted: *Alastor; or, the Spirit of Solitude*. . . . The poem treated
the spirit of solitude as a spirit of evil.'[21]

One can imagine Shelley giving assent to this in conversation
with Peacock, warmly approving, even supplying, a description of
'solitude' as 'a spirit of evil'. Yet the imagination is often drawn to
what the moral intelligence shies away from. *Alastor* immerses itself
and its readers in an experience which is valuably ambiguous and often
poignant in its 'mingling' of feelings, as Mary Shelley's thoughtful
summary brings out:

None of Shelley's poems is more characteristic than this. The solemn spirit
that reigns throughout, the worship of the majesty of nature, the broodings
of a poet's heart in solitude—the mingling of the exulting joy which the
various aspect of the visible universe inspires with the sad and struggling
pangs which human passion imparts—give a touching interest to the whole.
(*PW* 31.)

---

[21] 'Memoirs of Percy Bysshe Shelley', *The Works of Thomas Love Peacock*, The
Halliford Edition, ed. H. F. B. Brett-Smith and C. E. Jones (10 vols., London and
New York, 1924–34), viii (1934), 100.

2

'Hymn To Intellectual Beauty' and 'Mont Blanc': Peculiarly Emotional?

Alastor uses narrative as a vehicle for exploration of the psyche and is most impressive when avoiding overstatement. In two important shorter poems, 'Hymn to Intellectual Beauty' and 'Mont Blanc', written the same year as *Alastor* was published (1816), Shelley places his own consciousness at the centre of his poetry. Although the switch from narrative to lyric strengthens the poetry's subjective element, the presentation of experience in both shorter poems is intriguingly oblique. Indeed, the reader of Shelley often encounters a kind of poetry which both answers to and deserves to be defended against F. R. Leavis's hostile account: 'there is certainly a sense in which Shelley's poetry is peculiarly emotional, and when we try to define this sense we find ourselves invoking an absence of something'.[1] The comment is penetrating as description, unfair as adverse criticism. Shelley's presentation of emotion is capable of a reticent finesse, an ability to render its 'evanescent visitations' (*A Defence of Poetry*, PP 504) that Leavis's insistence on enactment fails to recognize. Francis Mulhern argues convincingly that Leavis saw the workings of Shelley's style as pointing towards 'a radical perceptual instability whose source was the devaluation of reason in favour of emotion'.[2] This chapter does not propose to celebrate any such 'devaluation', but to question the validity of the Leavisite distinction that Mulhern spells out. Shelley's poetry often makes the reader conscious of an affective way of experiencing which lies beyond and below the possibilities of argument. But this way of experiencing is intimately connected with the rational even as it serves as a signal of the poet's sceptical awareness of the limits of reason. If Shelley uses words in particular ways to make states of mind available, this making available combines with a strong sense, communicated by the poems, of the gap between words and feelings. The result is a poetry that is hard to evaluate, variable in quality, but rewarding in unexpected ways.

[1] Leavis, 194.   [2] *The Moment of 'Scrutiny'* (London, 1979), 133.

In the 'Hymn to Intellectual Beauty' and 'Mont Blanc', emotion comes in spurts and rushes, unpredictable offshoots; it is communicated by moments of split-second, tell-tale verbal excitement. Such moments usually express a temporary disruption, swerve, or subtilizing in the poem's processes of thought. Shifts and surfacings of emotion can seem almost inaccessible, and the reader may appear only to glimpse vanishing points beyond the poetry's absorption in speculative issues:

> Thy light alone—like mist o'er mountains driven,
>   Or music by the night wind sent
>   Through strings of some still instrument,
>   Or moonlight on a midnight stream,
> Gives grace and truth to life's unquiet dream.
>
> ('Hymn', 32-6)

>   the strange sleep
> Which when the voices of the desart fail
> Wraps all in its own deep eternity;—
>
> ('Mont Blanc', 27-9)

Yet in both passages 'emotion' manifests itself, though it is not offered, as Leavis has it, 'unattached, in the void';[3] in both cases the language subtly corroborates the sense of swerving suggested by the rhythms. 'Alone' in the first passage, 'strange' in the second, register change of mood most sensitively. Although the language is swift and unpausing, it repays attention, as is illustrated by the placing of 'fail' at the end of the line in the passage from 'Mont Blanc' to suggest the 'failing' of voice which Shelley has suddenly apprehended; or by the humanist adaptation of religious desire in the last line of the passage from the 'Hymn' ('grace and truth' occurs in John 1: 14). The blend of energy and forlornness in both passages is the consequence of the poet's interest in states of mind and self-created deities which he both wishes to bring into existence through language and fears may lie beyond words.

The passage from the 'Hymn' is set in what Spencer Hall calls 'radical juxtaposition'[4] to the scepticism about objects of traditional

---

[3] Leavis, 201.
[4] 'Power and the Poet: Religious Mythmaking in Shelley's "Hymn to Intellectual Beauty"', *KSJ* xxxii (1983), 143.

belief expressed in the preceding lines of stanza three. But the poetry's way of working is less self-announcing than 'radical juxtaposition' suggests.[5] Shelley's similes run at a suggestive tangent from the summarizing final line. 'Still instrument', for example, informs the conventional image of an Aeolian harp with unexpected feeling. The phrase haunts because the poet seems briefly to describe his own poem, an 'instrument' that is 'still' in the sense of having just grown responsive to (though not, as yet, fully possessed by) 'Thy light alone'. The way 'still' hints at something deeper is typically subliminal or, rather, liminal: the poetry poises us on a threshold, making us look before and after. Such moments are the more authentic for their stealthy arrivals and quick departures. The 'Mont Blanc' passage is similar in its effect. It is as if the lines sought at the same time to capture a state of perception which comes into being 'when the voices of the desert fail' and to preserve the elusiveness of this state where what is perceived is the inability to sustain sense perception. The poetry does not present us with emotions that are vague and 'unrealized' (to use the Leavisite word). Instead, it creates for us a sense of the difficulty of realizing certain emotions and gives us a glimpse of what it would be like to overcome such difficulty.

The subject of both poems is enigma ('the ultimate mysteriousness of man's subjective and existential being'),[6] and their tones are enigmatic. In the 'Hymn', Shelley constructs a faith in which to believe, while remaining fully aware of the extent to which his deity is capricious, arbitrary, a human projection. 'Mont Blanc' is agnostic rather than rhapsodic in spirit, its responsiveness to the sublime a sceptical, ambiguous affair. Judith Chernaik rightly contends that the truth which the poem 'defines is poetic rather than philosophical',[7] and what needs to be stressed is the trickily oblique relationship in both these poems between the poet's tackling of large themes and his discovery of a distinctive poetic voice.

The texture and impact of the poems of 1816 rely on the coexistence of passion and discipline. The poems glitter with a self-watchful care. For example, Shelley's reined-in formulation, 'the very spirit fails' ('Mont Blanc', 57), contrasts favourably with

---

[5] Hall does, however, mention the 'subtle imagistic and, ultimately, thematic shifts that it is all too easy to overlook but that provide inner tension in Shelley's best work', 'Power and the Poet', 137.
[6] 'Power and the Poet', 129.   [7] Chernaik, 46.

Coleridge's more expansive concession to the sublime in a probable source of 'Mont Blanc', the 'Hymn before Sun-rise, in the Vale of Chamouni', where he speaks of 'the dilating Soul' ('Chamouni', 21).[8] Even the couplet at the end of the fifth stanza of the 'Hymn'— 'Sudden, thy shadow fell on me; | I shrieked, and clasped my hands in extacy!' (59–60)—manages not to be compromised by its 'extacy' as Coleridge is by his 'swelling tears, | Mute thanks and secret ecstasy' ('Chamouni', 25–6). It is less accurate to regard the pitch of Shelley's lines as false than as justified dramatically, a repudiation of the inert preceding lines: 'When musing deeply on the lot | Of life, at that sweet time when winds are wooing | All vital things that wake to bring | News of buds and blossoming' (55–8). The disruption of the poem's texture is the reverse of reticent, yet it bears witness to an intensity of reminiscence that is barely tolerable and scarcely utterable. Such 'an emotionalism which finds no adequate translation into words'[9] arouses reservations and sympathy in almost equal measure. By contrast, Coleridge's modulations leave the reader unaffected, hardly ruffling the steady rapture sustained throughout his poem: his 'ecstasy' is the reverse of 'secret'.

Shelley's emotional effects are often most interesting when they are least declarative. When emotion does insist on recognition, in the fifth and sixth stanzas of the 'Hymn', the reader is tempted to deplore the presence of emphasis. The couplet already discussed at the end of the fifth stanza (59–60) wins, as suggested, a Pyrrhic poetic victory; its assertiveness betrays doubt; hectic alliteration tries to capture a feeling which obstinately refuses to be articulated. The sixth stanza is also damaged by insistence, and yet the writing struggles to search for subtler effects than its chosen details can convey. Shelley's invocation of 'the phantoms of a thousand hours', for instance, gestures towards innumerable epiphanic experiences, none

---

[8] In *SM* Harold Bloom asserts: 'Whatever the genetic history, the two poems are in violent contrast' (*SM* 12), but, for all his illuminating and pioneering comments on this contrast, he contends that 'there are no verbal echoes of the earlier "Hymn" in the later work' (*SM* 11). Angela Leighton, in my view correctly, argues in favour of a Coleridgean influence: 'There are a number of clear verbal echoes of the "Hymn" in "Mont Blanc", but the comparison is . . . more significant for the differences it brings to light' (Leighton, 58). As she points out, acknowledging Charles E. Robinson's *Shelley and Byron: The Snake and Eagle Wreathed in Fight* (Baltimore and London, 1976), Shelley or Byron almost certainly had a copy in Geneva of the 11th number of *The Friend* in which Coleridge's poem appeared.

[9] Leighton, 57.

of which he is able to realize in words, a lack of realization that
disappoints rather than intrigues:

> I vowed that I would dedicate my powers
>   To thee and thine—have I not kept the vow?
>   With beating heart and streaming eyes, even now
> I call the phantoms of a thousand hours
> Each from his voiceless grave . . .
>
> (61–5)

Here, the use of 'voiceless' fails to diagnose absence in the way that
makes 'voiceless lightning' (137) so impressive in the final section
of 'Mont Blanc' where the adjective earns its keep, suggesting the
poet's responsibility for any 'voice' that makes itself heard. The
formula of ineffability which rounds off the stanza reads as just
that: a formula that tries to extricate Shelley from the difficulty of
describing his deity's benefits. 'Dark slavery' abstractly politicizes a
predicament which has been intuited more sharply in preceding
stanzas (especially in stanza four with its fear that 'dark reality' (48)
may be an accurate description of existence):

> They know that never joy illumed my brow
>   Unlinked with hope that thou wouldst free
>   This world from its dark slavery,
>   That thou—O awful LOVELINESS,
> Wouldst give whate'er these words cannot express.
>
> (68–72)

Hard on the heels of the coy paradox—'O awful LOVELINESS'—the
final disclaimer trivializes the difficulties that words encounter in the
'Hymn'.

The opening lines glimpse these difficulties more persuasively:

> The awful shadow of some unseen Power
>   Floats though unseen amongst us,—visiting
>   This various world with as inconstant wing
>   As summer winds that creep from flower to flower.—
>
> (1–4)

Shelley's language combines the lucidly marshalled and the hoveringly
enigmatic. The reference to 'some unseen Power' braids the two
effects, prompting questions in the reader. Is this Power to be
identified with the 'Spirit of BEAUTY' (13) addressed in the second

stanza? Or is beauty merely one of its attributes? The poem itself does not define the relationship between 'Power' and 'Beauty'. An account of the 'Hymn' needs to acknowledge this evasiveness and to respect the intricate positions which the poem sustains: 'Intellectual Beauty' is addressed as though it were a deity, but the poem sees 'Beauty' as manifesting itself within 'human thought or form' (15); 'Beauty' is hailed as the source of meaning and value in this world, but is described as merely the 'awful shadow of some unseen Power'.

Though the poetry preserves the unknowability of the 'unseen Power', it hints at experience of its 'awful shadow'. If *some* unseen' (my italics) is deliberately unspecific, the ensuing phrase, 'though unseen', suggests a readiness to live with blur, a resolute commitment to intuition. Indeed, Shelley shows just as much relish for distinctions as doubt about the adequacy of his language. For instance, 'various world' slots into the niche that the philosophical idiom has created for it. The stanza also displays the willingness to speak in general terms which John Berryman stigmatizes as 'radar romanticism'.[10] Though Berryman means to be critical, his phrase points to the authority possessed by a couplet such as this: 'It visits with inconstant glance | Each human heart and countenance' (6-7).

Leavis's point—'we find ourselves invoking an absence of something'—is relevant to the gap between this firm assertion and the leap which the poetry then takes into a series of mockingly tenuous similes:

> Like hues and harmonies of evening,—
>   Like clouds in starlight widely spread,—
>   Like memory of music fled,—
>   Like aught that for its grace may be
> Dear, and yet dearer for its mystery.
>
>   (8-12)

What Shelley offers are comparisons which implicitly comment on their own figurativeness, underscoring the extent to which the poem's subject defeats language. To say this, however, is not necessarily an adequate retort to allegations of poetic weakness. Shelley comes as close to wry self-deprecation as he can, and what could be seen as his critique of the obsessive Wordsworthian image, the 'Immortality Ode''s single Tree, looks perilously slight, his phrasing dictated by

---

[10] 'Hardy and His Thrush', *The Freedom of the Poet* (New York, 1976), 244.

a facile reliance on alliterative links. Leavis, one imagines, would
have disliked this passage; his case against Shelley rests most firmly
on his belief that, by contrast with Wordsworth who 'seems always
to be presenting an object', the emotion seeming 'to derive from what
is presented', Shelley 'offers the emotion in itself'.[11] But the 'Hymn'
is written out of a sense of a gap as well as alliance between sensory
experience and evoked emotion, a gap which Shelley's similes both
try to bridge and acknowledge; the poem does not cede supremacy
to the sensory as the validating source of emotion, though it ranges
among sensory experience for evidence to support its intuitions and
desires.

   It is arguable, however, that this stanza's excursion into doubt and
exploration seems perfunctory. The swiftness with which Shelley
works can mean that issues or problems are left behind too quickly.
In stanza two, for example, the language has a winnowed feel as
the poetry flits between conflicting moods. Throughout the stanza,
Shelley uses traditional religious phrasing, but the very fluency of
the writing threatens to erase irony:

> Spirit of BEAUTY, that dost consecrate
>   With thine own hues all thou dost shine upon
>   Of human thought or form,—where art thou gone?
>
>   (13–15)

Certainly, 'consecrate' is adapted to a humanist context, as the
ensuing 'human thought or form' brings out. But Shelley's address
to the 'Spirit of BEAUTY' is virtually indistinguishable from orthodox
prayer. What follows is an example of the rapidity with which the
poetry changes emotional tack:

> Why dost thou pass away and leave our state,
> This dim vast vale of tears, vacant and desolate?
>   Ask why the sunlight not forever
>   Weaves rainbows o'er yon mountain river,
> Why aught should fail and fade that once is shewn,
>   Why fear and dream and death and birth
>   Cast on the daylight of this earth
>   Such gloom,—why man has such a scope
> For love and hate, despondency and hope?
>
>   (16–24)

---

[11] Leavis, 200, 201.

Lines 16–17 offer a miniature example of the tonal difficulties and rewards encountered in a reading of the 'Hymn'. Shelley laments that we live in a 'vale of tears', and the familiarity of the phrase cheats us into assent. But we recognize that the emotion has been quickly distanced; the ensuing questions discipline too facile an expression of grief, shrugging off sorrow as stereotyped and hackneyed. As a result we are left with a topos—life as a 'vale of tears'—which crumbles into a cliché. Only in 'vacant' does Shelley translate the predicament into a term which has more urgency. However, the correction of lines 16–17 by the following passage turns on itself arrestingly. The passage begins by accepting transience as inevitable, but its cumulative effect is to include 'Why aught should fail and fade that once is shewn' in a catalogue of mysteries which puzzle the poet, making him feel both scepticism and a desire for faith. At line 21 the poem sketches the shadowy forces which circle 'the daylight of this earth'; though Shelley's abstract shorthand—'love and hate, despondency and hope'—curtails the poetry's resonance, his sense of the numinous is haunting.

Throughout stanza two, Shelley's language seems prompted by a wish not to go over the same ground as Wordsworth's 'Immortality Ode'. Harold Bloom is surely right to see Wordsworth's poem as a model for Shelley's, but wrong to assert that the 'Hymn' 'barely deviates' from the earlier poem.[12] The 'Hymn' could be said to shrink from a poignancy which it does not want to emulate. For example, Wordsworth's striking transitions give way to Shelley's scarcely noticeable changes of mood; and Wordsworth's elegiac questions—'Whither is fled the visionary gleam? | Where is it now, the glory and the dream?' (56–7)—are rationalized as inevitable consequences of transience: 'Why aught should fade and fail that once is shewn . . . '. (That said, Shelley smuggles back, as already suggested, the regret which this line appears to outlaw.) Even when the debt of the 'Hymn' to the 'Immortality Ode' is most blatant, its tones have an individual persuasiveness. Though the final stanza's mood of acceptance has more than a tinge of Wordsworth's 'sober colouring' ('Immortality Ode', 201)—

[12] *A Map of Misreading* (1975; Oxford, New York, Toronto, and Melbourne, 1980), 149.

> The day becomes more solemn and serene
>   When noon is past—there is a harmony
>   In autumn, and a lustre in its sky,
>   Which through the summer is not heard or seen,
>   As if it could not be, as if it had not been!
>
>   (73–7)

—Shelley suggests the limits of scepticism with his own brand of teasing mockery. The lines accelerate high-spiritedly as the poet repudiates earlier instances of the rational obstinacies of eye and ear, especially those at line 30 (where Shelley cannot separate 'From all we hear and all we see, | Doubt, chance, and mutability') and at line 54 (where Shelley writes 'I was not heard—I saw them not—'). Now he senses a harmony and lustre 'Which through the summer is not heard or seen, | As if it could not be, as if it had not been!', where the last line's cadence surmounts and controls the pathos which its original in the 'Immortality Ode' possesses ('It is not now as it hath been of yore' (6)).

Appropriately the concluding lines of the 'Hymn' are cast in the form of a prayer:

> Thus let thy power, which like the truth
>   Of nature on my passive youth
> Descended, to my onward life supply
>   Its calm—to one who worships thee,
>   And every form containing thee,
>   Whom, SPIRIT fair, thy spells did bind
> To fear himself, and love all human kind.
>
>   (78–84)

Here, 'power' is seen as an attribute of 'SPIRIT', not as the ultimate principle of the Universe, a reversal of the poem's opening position that suggests the obedience of Shelley's ideas to his feelings. Having established a relationship with Intellectual Beauty, he can now regard its 'power' as available to him through prayer. There is a sleight-of-hand about the ending, the speed of whose rhythms indicates an impatient desire to affirm which is affectingly at odds with the poet's request for 'calm'. 'Hymn to Intellectual Beauty' sustains its fluctuations of feeling to the end.

Central to the achievement of the 'Hymn' and 'Mont Blanc' is the capacity to make discoveries. In stanza three of the 'Hymn' Shelley begins with a generalization which he goes on to qualify and complicate:

No voice from some sublimer world hath ever
  To sage or poet these responses given—
  Therefore the name of God and ghosts and Heaven
Remain the records of their vain endeavour,
Frail spells—whose uttered charm might not avail to sever,
    From all we hear and all we see,
    Doubt, chance, and mutability.
Thy light alone . . .

(25–32)

The unspecificity of 'some sublimer world' casts doubt on the existence of any such world, but it also allows the poet room for manœuvre. He may dismiss with agnostic contempt calls upon 'the name of God and ghosts and Heaven' as so many 'Frail spells'. In his subsequent address, however, Shelley embarks on his own 'frail spell' as, joining scepticism and faith, intelligence and passion, he invokes 'Thy light alone'. The unadvertised swiftness of the transition, as argued at the start of the chapter, is characteristic and, in this case, highly effective.

In stanza four Shelley describes the cycle of change from which he claims Intellectual Beauty could redeem him. But awareness of transience moves abruptly into fear by way of another simile:

Thou—that to human thought art nourishment,
  Like darkness to a dying flame!
  Depart not as thy shadow came,
  Depart not—lest the grave should be,
Like life and fear, a dark reality.

(44–8)

The paradox of 'Like darkness to a dying flame' exposes Shelley's awareness that his invocation of a 'Thou' may be falsifying and absurd. 'Dark reality' uncovers the fear that the 'grave' marks an absolute limit. Yet the passage retains its equilibrium; its terrors are couched in a lyrical form that suggests the poet's control, a counterbalancing typical of a poem which is clear-sighted yet open-minded, and dramatizes successfully the clash between impulses of uncertainty and faith.

In *The Unmediated Vision* Geoffrey Hartman writes: '*Wordsworth's understanding is characterized by the general absence of the will to attain relational knowledge*, that is, knowledge which may be obtained in direct answer to the Why, the What, the Wherefore,

Such a *will to attain relational knowledge* is both
jected to scrutiny in 'Mont Blanc', a poem which
on, even if it does not wholly exorcize, the need
..ypothesize answers. As Gerald McNiece remarks, 'Mont
...nc' 'demonstrates the value of poetry as a way of knowing
relatively free from the passion for secure and stable knowledge'.[14]
But Shelley's self-consciousness is not an automatic censoring-device
that holds him aloof from belief. Rather, it gives rise to variations
of tone and quality which the poem's very energy may seem to
disguise. As William Keach has demonstrated, waywardness is built
into 'Mont Blanc' by virtue of its irregular rhyming, 'both a stay
against and a means of marking the chaos and blankness which are
Mont Blanc's special concerns'.[15]

The opening section raises the problem of the poem's dealings with
'philosophy':

> The everlasting universe of things
> Flows through the mind, and rolls its rapid waves,
> Now dark—now glittering—now reflecting gloom—
> Now lending splendour, where from secret springs
> The source of human thought its tribute brings
> Of waters,—with a sound but half its own.
>
> (1–6)

Shelley's phrasing has the ring of someone setting out an initial
premiss: 'The everlasting universe of things | Flows through the mind
. . . '. The lines may seem to restate an epistemological commonplace.
Yet if the excited effect of 'everlasting' brings Eliot's half-reproving
comment on Shelley to mind—'abstractions could excite in him
strong emotion'[16]—the fact that the poem's energies are not wholly
at one with its ratiocinative ambitions is a strength. There is a
swift, self-aware interplay in 'Mont Blanc' between the poet and his
creation which is inadequately described by Leavis as 'confusion';
he disapproves of the way that the 'metaphorical and the actual, the
real and the imagined, the inner and the outer, could hardly be more

---

[13] *The Unmediated Vision: An Interpretation of Wordsworth, Hopkins, Rilke, and Valéry* (1954; New York, 1966), 5.
[14] 'The Poet as Ironist in "Mont Blanc" and "Hymn to Intellectual Beauty" ', *SIR* xiv (1975), 326.
[15] Keach, 196.
[16] 'Shelley', *Selected Prose of T. S. Eliot*, ed. Frank Kermode (London, 1975), 82.

unsortable'.[17] But there is power in Shelley's evocation of a state in which distinctions fail.

What is valuable about the opening is the way the impulse to categorize glimpses its own limitations and becomes unresisted vertigo. Almost by chance, the 'feeble brook' (7), Shelley's image for 'The source of human thought' (5), slips its emblematic leash, finds a life of its own:

> Such as a feeble brook will oft assume
> In the wild woods, among the mountains lone,
> Where waterfalls around it leap for ever,
> Where woods and winds contend, and a vast river
> Over its rocks ceaselessly bursts and raves.

> (7–11)

The writing restores reality to the physical, directing attention to a landscape that we momentarily apprehend as actual. This comes as a surprise after the subjection of the physical to the speculative in the opening passage. And yet the lines survive as a way of talking about the mind since, for all the writing's physicality, the generic features of the landscape imply the mind abstracting and arranging experience. The imaginative impact of the poetry derives from its refusal to support an argument or to pin its faith on concrete sensuousness.

Physical realities make their presence felt in the second section:

> Thus thou, Ravine of Arve—dark, deep Ravine—
> Thou many-coloured, many-voiced vale,
> Over whose pines, and crags, and caverns sail
> Fast cloud shadows and sunbeams . . .

> (12–15)

But the poetry belies its promise of point-by-point comparison. Indeed, as Wasserman points out, 'The truncation of the simile ["Thus thou, Ravine of Arve"] . . . is apposite',[18] though not, *pace* Wasserman, because the writing deliberately fails to take full account of the 'intellectual philosophy', but because it is engaged in probing, questioning. Attention to the scene's correspondences with Shelley's speculative terms in the opening passage is twice interrupted.

[17] Leavis, 199.   [18] Wasserman, 226.

The first instance occurs when Shelley glosses the actual as an 'awful scene, | Where Power in likeness of the Arve comes down | From the ice gulphs that gird his secret throne, | Bursting through these dark mountains . . . ' (15–18), lines which bring out the poet's use of the 'awful scene' as a dramatization of his 'imaginings' about some postulated 'Power', and make us aware of the poem as an artifice, something constructed. Shelley explores the implications of his intuition, recorded in a letter written after his journey to the Valley of Chamouni, that 'All was as much our own as if we had been the creators of such impressions in the minds of others, as now occupied our own'.[19] Subsequent lines mingle this awareness of artifice with a recognition of natural 'harmony':

> thou dost lie,
> Thy giant brood of pines around thee clinging,
> Children of elder time, in whose devotion
> The chainless winds still come and ever came
> To drink their odours, and their mighty swinging
> To hear—an old and solemn harmony . . .
>
> (19–24)

Here, the religious phrasing both relates to and, by virtue of its metaphoric boldness, contends with the claim which immediately follows in the letter cited above: 'Nature was the poet whose harmony held our spirits more breathless than that of the divinest.' At this stage the religious feeling of the poem is a tentative way of putting things, a fanciful prelude to sterner investigations: the relationship between 'Nature' and the 'human mind's imaginings' is left finely poised. Judith Chernaik believes that 'The ravine suggests a natural religion',[20] and Shelley's use of religious language could be regarded as containing an element of grave parody.

The second moment when the poet departs from the actual during this section takes place in the lines describing 'Thine earthly rainbows stretched across the sweep | Of the etherial waterfall, whose veil | Robes some unsculptured image' (25–7). There is a tone, here, of half-stifled rapture, although the progression from 'earthly' to 'etherial' and then to 'unsculptured' is too deliberate for 'some unsculptured image' to act as an unsettling revelation of the poet's own activity. Still, 'unsculptured' equivocates between suggesting the

---

[19] *Letters*, i. 497.   [20] Chernaik, 43.

aloofness of the 'image' from the physical landscape and implying the poet's presence in the act of 'unsculpturing' or decreating the 'real'.

Though 'Mont Blanc' continually edges towards an assertion of the mind's creative powers, such an assertion is balanced by growing anxiety about whether 'things' can be known. Sensuous loss may lead to imaginative plenitude. But the poet is haunted by the idea that 'eternity' is experienced in a limbo state, a 'strange sleep | Which when the voices of the desert fail | Wraps all in its own deep eternity'. Shelley's fascination with his own mental experience only fully emerges at line 34. Before this, it is the surplus energy of the poetry which hints at the presence of more than physical description. The evocation of the 'Arve's commotion' is an example:

> Thy caverns echoing to the Arve's commotion,
> A loud, lone sound no other sound can tame;
> Thou art pervaded with that ceaseless motion,
> Thou art the path of that unresting sound—
>
> (30–3)

Rhythmically these lines duplicate 'A loud, lone sound no other sound can tame', offering what, after the 'strange sleep', comes across as consoling access to the physical. Yet the word 'lone' links such access to a disconcerting memory of an earlier solitude, the half-real 'mountains lone' (8) of the opening section, while 'ceaseless' and 'unresting', far from being verbal driftwood, are troubled by the prospect of 'cessation' and 'rest'.

So far, then, the poetry has intuited uncanny gleams on the far side of its epistemological concerns. Harold Bloom pinpoints a central area of debate when he writes: 'A close reading of the second section . . . ought not to find secondhand philosophy, but rather a firsthand account of a personal religious experience, an experience centered in the composition of poetry.'[21] Yet the poetry swings between the imaginative and the rational, each mode putting pressure on the other. At times it does seem to verge on the 'secondhand philosophy' which Bloom outlaws from critical discussion:

> Dizzy Ravine! and when I gaze on thee
> I seem as in a trance sublime and strange
> To muse on my own separate phantasy,
> My own, my human mind, which passively

[21] SM 30.

> Now renders and receives fast influencings,
> Holding an unremitting interchange
> With the clear universe of things around . . .
>
> (34–40)

Both language and rhythm, here, are cogently suggestive of the toing and froing which form 'interchange'. But for Shelley to describe his mind as engaged in 'an unremitting interchange | With the clear universe of things around' is to provoke philosophical paraphrase. The passage teeters between stating epistemological truisms and trying to evoke 'a trance sublime and strange'. Though Shelley's painstakingly appositional phrasing, 'my own separate phantasy, | My own, my human mind', may rouse sympathy, the writing labours its insights when it draws attention to the use of the Ravine as an emblem for the processes of consciousness. What prevents this from being wholly the case is that the Ravine is object as well as emblem of the poet's thinking. In its otherness it prompts in Shelley the discovery that mind is distinct from object, that the object can never be fully drawn up into the mind, that language can only stand in for objects. It is the rich fusion and confusion of intuition and thought that give value to such passages as the one just quoted and the one which immediately follows:

> One legion of wild thoughts, whose wandering wings
> Now float above thy darkness, and now rest
> Where that or thou art no unbidden guest,
> In the still cave of the witch Poesy,
> Seeking among the shadows that pass by
> Ghosts of all things that are, some shade of thee,
> Some phantom, some faint image; till the breast
> From which they fled recalls them, thou art there!
>
> (41–8)

Shelley's language enacts his fascination with the indeterminate relationship between mind and world. This fascination gives rise to the paradox of lines 37–8, where he presents his own mind as '*passively*' (my italics) rendering as well as receiving 'fast influencings', as if the mind were, in part, not the creator of the thoughts that it sends beyond itself. The syntax of lines 41–8 is even trickier, and the passage offers a good example of the evaluative issues thrown up in the wake of interpretative cruces. At least four difficulties offer evidence not of bad writing but of Shelley's struggle to realize elusive experience: (1) the antecedent of 'Seeking' in line 45; (2) the possessor of the 'breast' in line 47; (3) the identity of 'they' in line 48; (4) the identity of 'thou' in the same line. That so much of the poetry is hard to assign a meaning is indicative of the tussle between the wish to state and the desire to evoke in this section of the poem.

That said, I would supply the following answers to the difficulties outlined above: (1) the antecedent of 'Seeking' is hard to locate— the 'legion of wild thoughts' is the most likely candidate—but the unspecificity expresses the poet's uncertain view of his own mind as a subject which engages in 'Seeking'; the poem sees the mind as the sum of its thoughts about things which can never be grasped in themselves, but only known through images; (2) the attempt to solve this difficulty provokes the most crucial ambiguity in the passage, since the 'breast' of line 47 can be seen as belonging to an entity beyond the human mind (on this reading the human mind is equivalent to the 'legion of wild thoughts' lent by some 'Power', a reading which acknowledges the tenuous hold maintained by Shelley on the separate identities of 'self', 'Power', and 'reality'); on the other hand, the 'breast' can also be seen, with Bloom, as 'that of Shelley';[22] (3) the identity of 'they' in line 48 could be either the 'legion of wild thoughts' or the 'shadows' of line 45, though the second reading is preferable because the 'shadows' relate specifically to the process of imaginative revelation traced in the last four lines where 'shadows' are the poetic means by which Shelley strives to embody his ungraspable concept of 'Power'; (4) that last phrase, 'ungraspable concept of "Power"', explains the identity of 'thou' (48), a word that has advanced beyond its meaning at line 43 in which it refers to the Ravine (performing what Wasserman calls a 'synecdochic role as the universe');[23] in the final words, 'thou art there!', 'thou' emerges as something discovered, glimpsed in its uniqueness at the moment of its disappearance from language, salvageable only in the form of a brief cry.

The passage manages, then, to offer the reader more than 'second-hand philosophy'; it involves us in the struggle to capture intuitions and thoughts as they come into consciousness; it ends up as being itself an experience rather than a statement about experience. The lines bring out the need to stress the poetic nature of Shelley's dealings with ideas. One can measure the strengths and limits of Shelley's

22 *SM* 30.   23 Wasserman, 227.

speculative idiom by comparing it with Coleridge's use of language in 'The Eolian Harp'. Arguably, Shelley never 'thinks' in his verse with the authority or clarity of Coleridge's pantheist surmise: 'And what if all of animated nature | Be but organic Harps diversely fram'd . . . ?' (44–5), but he mimes the experience of thought with greater immediacy. Coleridge's 'Abstruser musings' ('Frost at Midnight', 6) rarely fail to discover a lucid image or tidy epigram. In 'The Eolian Harp', for example, the passage cited above comes to rest on this note: 'At once the Soul of each, and God of all' (48). Lacking such certainty, Shelley's phrasing in 'Mont Blanc' substitutes its own intensity. 'Ghosts', for instance, is a far more active register of internal states than one finds in Coleridge. A similar conclusion emerges if one contrasts Shelley's lines in the second section and a passage from Coleridge's poem which may have left its imprint on 'Mont Blanc':

> Full many a thought uncall'd and undetain'd,
> And many idle flitting phantasies,
> Traverse my indolent and passive brain,
> As wild and various as the random gales
> That swell and flutter on this subject Lute!
>
>   (39–43)

One parallel in 'Mont Blanc' to these lines occurs in Shelley's decision to 'muse on my own separate phantasy'; possibly drawn from Coleridge's 'many idle flitting phantasies', the phrase has become far more inward in Shelley's poem. If Shelley's passage evokes 'fast influencings', the effect of Coleridge's lines is one of relaxed subtlety.

Between sections two and five, however, comes the poet's attempt to spell out the polemical and political meanings of his experience. In effect, Shelley composes a sceptical reply to Coleridge's challenge: 'Who *would* be, who *could* be an Atheist in this valley of wonders!'[24] The lines which 'Mont Blanc' could be said to contest most sharply are these:

> Around thee and above
> Deep is the air and dark, substantial, black,
> An ebon mass: methinks thou piercest it,
> As with a wedge! But when I look again,
> It is thine own calm home, thy crystal shrine,
> Thy habitation from eternity!

[24] *Coleridge: Poetical Works*, Oxford Standard Authors, ed. E. H. Coleridge (1912; London, Oxford, and New York, 1969), 377.

O dread and silent Mount! I gazed upon thee,
Till thou, still present to the bodily sense,
Didst vanish from my thought: entranced in prayer,
I worshipped the Invisible alone.

('Chamouni', 7–16)

Although the seductions of a trance-like gaze ascending out of thought are experienced elsewhere in Shelley, the reader is more conscious in 'Mont Blanc' of an alert scepticism. In a letter Shelley was less wary about the intensity of his experience than in the poem, writing: 'I never knew I never imagined what mountains were before. The immensity of these aerial summits excited, when they suddenly burst upon the sight, a sentiment of extatic wonder, not unallied to madness.'[25] The poem stays at a remove both from this 'madness' and from Coleridge's rapture as it moves into the self-questioning third section:

Some say that gleams of a remoter world
Visit the soul in sleep,—that death is slumber,
And that its shapes the busy thoughts outnumber
Of those who wake and live.—I look on high . . .

(49–52)

Shelley merely reports what 'Some say', even if the appeal of those sonorous 'gleams of a remoter world' is evident. Yet where Coleridge's second glance, 'But when I look again' ('Chamouni', 10), is radiant with certainty—'It is thine own calm home' ('Chamouni', 11)— Shelley's accent is virtually toneless as 'I look on high' addresses dangerously large issues. It is a strength of the poetry that its hesitations show awareness of such dangerous largeness. The placing of 'inaccessibly' at the end of a line is a case in point, the word at once retreating from and spinning out speculation:

Has some unknown omnipotence unfurled
The veil of life and death? or do I lie
In dream, and does the mightier world of sleep
Spread far around and inaccessibly
Its circles?

(53–7)

[25] *Letters*, i. 497.

Working with familiar abstractions, Shelley's language teases us into thought. The precise meaning of these lines hinges on 'unfurled'. Murray's view that Shelley meant exactly what he wrote (that the veil is let down and not 'upfurled') seems the most plausible.[26] On this reading the poet draws a watchful distinction: either we are confined to a scepticism which an 'unknown omnipotence' renders complete, or the lines posit a 'mightier world of sleep' which may, if the section's opening guess is conceded, bring supra-rational intimations. These speculations are conveyed through a use of words that is subtle and reticent.

The powerful passage which follows sustains the balance between an awareness of limits and a sense of possibilities:

> For the very spirit fails,
> Driven like a homeless cloud from steep to steep
> That vanishes among the viewless gales!
> Far, far above, piercing the infinite sky,
> Mont Blanc appears,—still, snowy, and serene—
> Its subject mountains their unearthly forms
> Pile around it, ice and rock . . .
>
> (57–63)

The tone modulates significantly in these lines from a feeling of helplessness before the indifference of the universe to steadfast recognition of Mont Blanc as beyond human assimilation, 'still, snowy, and serene', no habitation of a traditional deity. But out of his sense of not being at home in the world, Shelley arrives at the most positive statement in the poem so far: the very silence evokes a voice, 'a mysterious tongue | Which teaches awful doubt, or faith so mild, | So solemn, so serene, that man may be | But for such faith with nature reconciled' (76–9). In the confusing phrase, 'But for such faith',[27] the poet seems to be led astray by an impulse to qualify which otherwise serves him admirably in the section (as in lines 71–4

---

[26] 'Mont Blanc's Unfurled Veil', *K S J* xviii (1969). Murray argues that Shelley used 'unfurled ' to 'refer . . . to the aftermath of the ecstasy, the *anti*-climactic moment when . . . the images of all things that are were called back by the power that had caused them to flow through the witch's cave', pp. 40–1.

[27] 'In such a faith', the reading in the fair copy of the poem in the Scrope Davies Find, irons out the obscure tension between man, faith, and reconciliation with nature which hovers around the lines I have quoted from *PP*. See Judith Chernaik and Timothy Burnett, 'The Byron and Shelley Notebooks in the Scrope Davies Find', *Review of English Studies*, new series, xxix (1978), 47.

which offer in the form of questions alternative accounts of the origin of the mountainous landscape).

Most impressive about sections three and four are their moments of quiet, pauses in which Shelley re-explores notions of 'silence', a silence which gives rise to but lies below the message his romantic ventriloquism conjures out of the landscape. Examples are offered by lines 57–63, quoted above, or by the account of 'A desert peopled by the storms alone' (67), or by the wary half-line, 'all seems eternal now' (75). Such evocation of silence is more compelling than the 'voice' which the poet coaxes from it, though put beside Coleridge's assertion—'And they too have a voice, yon piles of snow, | And in their perilous fall shall thunder, GOD!' ('Chamouni', 62–3)—Shelley's voice, 'not understood | By all, but which the wise, and great, and good | Interpret, or make felt, or deeply feel' (81–3), is cautious and tentative as the poet places responsibility for 'interpretation' firmly on the shoulders of human beings.

Intuition shades towards doctrine in the fourth section, where Shelley seems intent on a dignified but prosaic vigour, using words which have been important earlier, such as 'trance', 'serene', 'roll', 'for ever', and 'gleam', yet failing to enrich or refine their meanings. But the case alters in the final section when the poet's 'adverting mind' (100), anxious to spell out significance, gives way to a side of his sensibility which is more tactful and intense:

> Mont Blanc yet gleams on high:—the power is there,
> The still and solemn power of many sights,
> And many sounds, and much of life and death.
> In the calm darkness of the moonless nights,
> In the lone glare of day, the snows descend
> Upon that Mountain; none beholds them there,
> Nor when the flakes burn in the sinking sun,
> Or the star-beams dart through them:—Winds contend
> Silently there, and heap the snow with breath
> Rapid and strong, but silently! Its home
> The voiceless lightning in these solitudes
> Keeps innocently, and like vapour broods
> Over the snow. The secret strength of things
> Which governs thought, and to the infinite dome
> Of heaven is as a law, inhabits thee!
> And what wert thou, and earth, and stars, and sea,

> If to the human mind's imaginings
> Silence and solitude were vacancy?
>
> (127–144)

Here Shelley's ironic version of sublimity has little to do with conventional awe. A more adequate gloss on the poetic life of his imaginings is offered by a passage in Schiller's essay, *On the Sublime*:

It is precisely the entire absence of a purposive bond among this press of appearances by which they are rendered unencompassable and useless for the understanding (which is obliged to adhere to this kind of bond) that makes them an all the more striking image for pure reason, which finds in just this wild incoherence of nature the depiction of her own independence of natural conditions.[28]

The distinction between 'understanding' and 'pure reason' codifies what in 'Mont Blanc' is a more imaginative uncovering of the mind's awareness of its own activities. Still, the subtlety with which Shelley moves beyond 'the entire absence of a purposive bond among this press of appearances' is responsible for the finest single passage of poetry so far in his career. The opening lines of the section gather up the central concerns of the poem. Though their reassuring doubling, 'the power is there, | The still and solemn power of many sights, | And many sounds', risks being a shade too relaxed, the lines also bring out the equivocal status of 'power', the word flickering between suggesting the deified 'Power' of the fourth section and acting as a collective noun for the poet's acts of intuition.

In the next eight lines, however, the poetry breaks through to a more fascinating region. Shelley imagines what conditions are like at the top of the mountain; his imaginings convert stark blankness into a world of 'Silence and solitude'. Metaphysical implications inhere in Shelley's use of physical imagery to imagine what he, cannot see, implying the dependence of meaning on the mind. The evocation of the poet's simultaneous presence and absence is delicately managed: 'none beholds them there, | Nor when the flakes burn in the sinking sun, | Or the star-beams dart through them'. The opening disclaimer ushers in a ghostly play of appearances; the burning flakes and darting star-beams exist beyond the senses. Shelley's language uncoercively invites attention as it recedes into

---

[28] *'Naive and Sentimental Poetry' and 'On the Sublime'*, trans. Julius A. Elias (1966; New York, 1975), 205–6.

more intimate areas of consciousness: 'In the calm darkness of the moonless nights, | In the lone glare of day, the snows descend | Upon that Mountain'. The emphasis given to 'Upon that Mountain' relates the non-human absolute to the presence of the 'snows', but calls to mind the human consciousness engaged in charting the relationship.

In lines 134–6 Shelley may well rework lines from Coleridge's 'Chamouni' to create his most secretly exhilarated imagining in the poems of 1816. 'Chamouni' declares, 'but thou, most awful Form! | Risest from forth thy silent sea of pines, | How silently!' (5–7),[29] and, for once, believer and atheist share a common vision, each glimpsing a numinous 'silence' that repudiates the contention of sense-impressions in which 'Mont Blanc' is so rich. Shelley's repetition, 'but silently', swoops down, intent, on the state, prolonging its imaginative life. 'Solitude' is also explored and realized in the following lines. After this play of imagination, the claim which Shelley arrives at—his assertion that there is a 'secret strength of things' which governs thought and matter—is ready to contradict itself, so that thought is surmised to be the governor of the 'secret strength of things'. If the controversial final question does bring about such a contradiction, it also recognizes the tenuous stay which the mind's 'imaginings' provide against 'vacancy'. Like the 'Hymn to Intellectual Beauty', 'Mont Blanc' thrives imaginatively on switches of direction; it concludes with one that is unemphatic but resonant.

[29] Jonathan Wordsworth asserts: 'There can be no doubt that he [Shelley] is specifically recalling the use of repetition in *Hymn Before Sunrise*', 'The Secret Strength of Things', *The Wordsworth Circle*, xviii (1987), 107.

3

Julian and Maddalo
Privacy and Revelation

Two years after the composition of the 'Hymn to Intellectual Beauty' and 'Mont Blanc', Shelley began *Julian and Maddalo*, a poem which creates a new form for, and builds on, the instinct to dramatize inner conflict evident in the poems of 1816. The result is a crucial poem in Shelley's career. *Julian and Maddalo* is subtitled *'A Conversation'*. Yet its epigraph from Virgil's *Eclogue* X ('The meadows with fresh streams, the bees with thyme, | The goats with the green leaves of budding spring, | Are saturated not—nor Love with tears') is highly wrought, posing an immediate challenge to what Donald Davie calls the 'decorum of the conversation piece'.[1] The juxtaposition of subtitle and epigraph prepares the reader for the poem's exploration of tensions between different ways of registering experience. Certainly, the passionate excess of the Maniac's soliloquy bears witness to the complexity with which Shelley contextualizes his use of a familiar style in *Julian and Maddalo*. It is a style which both frames and is framed. In a recent essay Kelvin Everest makes an ambitious bid to explain the purpose of this framing; he contends that the Maniac's mode of utterance represents an ideal of 'critically disruptive emotional engagement with the conflict between social aspiration and social reality',[2] and that Julian and Maddalo speak a language which is tainted by the prejudices of a dominant class. The idea that one can discover in Shelley's work the same 'nagging contradiction between manner and commitment'[3] that Everest finds in the life is interesting. But to see Shelley as engaged in a guilt-ridden 'placing' of 'false consciousness' is to unify the parts of *Julian and Maddalo* at the cost of the poem. It involves, as my argument will imply, an unusefully hostile stance towards some of the finest poetry in the

---

[1] Davie, 143.
[2] 'Shelley's Doubles: An Approach to *Julian and Maddalo*', *Shelley Revalued: Essays from the Gregynog Conference*, ed. Kelvin Everest (Leicester, 1983), 81.
[3] 'Shelley's Doubles', 64.

work (the opening, say) as illustrating 'the ways in which the repressive ideology of the State may operate as an internal function of consciousness'.[4] It means pouncing on Julian's ineffectuality, his inability to put into practice 'dreams of baseless good' (578), in a more accusatory spirit than the poem will allow.

However, *Julian and Maddalo* is a text that prompts the reader to supply explanations and moralizing glosses, even as it refuses to give final assent to any single account. In a letter to his publisher, Shelley hints at the friction between the quotidian and the intense to which 'a *sermo pedestris* way of treating human nature' might give rise: 'It is an attempt in a different style, in which I am not yet sure of myself, a *sermo pedestris* way of treating human nature quite opposed to the idealism of that drama [that is, *Prometheus Unbound*].'[5] As Richard Cronin points out, '*sermo pedestris*' is a description associated in the eighteenth century with 'any colloquial style that seemed to derive from Horace's manner in his satires or *sermones*';[6] the poem's attention to the interplay between two participants in a conversation is its way of pursuing the psychological ambitions also evident in *Alastor* and *Prince Athanase*. Shelley's Byronic absorption in Athanase's 'secret pain' (85, *PW* 161) induces a mixture of analysis and reticence, which is curious rather than satisfying. His reason for giving up *Prince Athanase* reveals an author fatally unable to decide how far he should go with his psychological probing: 'The Author was pursuing a fuller development of the ideal character of Athanase, when it struck him that in an attempt at extreme refinement and analysis, his conceptions might be betrayed into the assuming a morbid character.'[7] In turn, *Julian and Maddalo* provokes the reader to ask whether its 'extreme refinement and analysis' is humanely searching or merely indecisive.

---

[4] 'Shelley's Doubles', 67.   [5] *Letters*, ii. 196.   [6] Cronin, 109.
[7] *PW* 161. For evidence that the poem that Shelley sent to Ollier on 23 Dec. 1819 was called 'Athanase: A Fragment' and not *Prince Athanase* see vol. vii of *Shelley and his Circle 1773–1822*, ed. Donald H. Reiman (Cambridge, Mass., 1986), 110–11. 'Prince Athanase', the longer draft version from which the text of 'Athanase: A Fragment' derives, is tentatively dated 'between July 1818 and August 1819', *Shelley and his Circle 1773–1822*, vii. 151. See the same vol. for the distinction between *Prince Athanase* (the published version in *Posthumous Poems* and later edns.) and 'Prince Athanase' (the notebook draft), p. 144; and for the view that 'both the maniac's self-revealing outpourings in *Julian and Maddalo* and "Athanase: A Fragment" were composed in the second half of 1819', p. 111.

The Preface initiates this enquiry; it begins with the divided if clear-cut attitudes induced by Maddalo, 'a person of the most consummate genius' whose 'weakness' it is 'to be proud' (*PP* 112). But 'proud' is a word which the author of the Preface feels needs to be qualified as soon as it has been written: 'I say that Maddalo is proud, because I can find no other word to express the concentered and impatient feelings which consume him; but it is on his own hopes and affections only that he seems to trample, for in social life no human being can be more gentle, patient, and unassuming than Maddalo.'(*PP* 113.) The awareness that human beings resist simple definition makes a major contribution both to the poem's overall achievement and to its occasional opacity or elusiveness. In the next paragraph there is a quick recoil from making fun of the 'pious reader' (who is left to determine whether it is possible that the heterodox Julian possesses 'some good qualities') to mockery of Julian as 'rather serious' (*PP* 113). The effect is straightaway to distance Shelley from a character that the reader of, say, *Queen Mab* might suppose would have his unequivocal approval: Julian is a good Godwinian, 'passionately attached to those philosophical notions which assert the power of man over his own mind' (*PP* 113). Finally, Shelley asserts the Maniac's universal relevance in a formulation which borders on paradox by implying that the crucial 'text' is not the one we are about to embark on but the private 'heart' of every reader: 'His story, told at length, might be like many other stories of the same kind: the unconnected exclamations of his agony will perhaps be found a sufficient comment for the text of every heart.' (*PP* 113.) A crucial element of the central section (more fully discussed below) is that it offers revelations of almost confessional intensity yet respects, or is unable to do without, the notion of some ultimate privacy. In an attempt to claim generalized significance for its psychological soundings, Shelley's 'text' presents itself as a series of 'comments' on a prior, unarticulated 'text' that is both individual and collective: the 'text of every heart'.

For all its air of quiet authority, then, the Preface is alive to uncertainties. Central to the impact of the poem proper is the way Shelley leaves the reader unsure from time to time whether uncertainties are seen fully by the poet or whether they are glimpsed out of the corner of his eye. The poem's detachment means that it is able to avoid propaganda on behalf of any one version of experience; does it also mean that Shelley can choose to duck or waive judgements? *Julian and Maddalo* is preoccupied with competing impulses—especially the friction between the desire to communicate and the wish to keep secret—and is a poem whose capacity to reward is inseparable from its refusal to unperplex.

Many readers prefer those passages of *Julian and Maddalo* in which Shelley is aware, to use Donald Davie's phrasing, that 'accuracy confers its own dignity'.[8] And yet even the opening and concluding movements of the poem are troubled (and enriched) by feelings more complex than the quest for descriptive accuracy. In *Alastor's* evocation of 'a wide and melancholy waste | Of putrid marshes' (273–4) Shelley anticipates his opening's use of a desolate landscape he might have modelled on Crabbe's *Peter Grimes*. However, despite resemblances between details—Crabbe's 'blighted tree' (174) and 'entangled weeds' (179) may lie behind Shelley's 'dwarf tree' (10) and 'amphibious weeds' (5)[9]—the two descriptions achieve quite different effects. Crabbe's famous lines, caricatured by Hazlitt as 'an exact *facsimile* of some of the most unlovely parts of the creation',[10] carefully erase any gap between Peter's state of mind and the dreary scene; his rhythms take their cue from the monotony of this line: 'At the same time the same dull views to see' (173). Shelley, on the other hand, implies both Julian's excited participation in and distance from the natural:

> I rode one evening with Count Maddalo
> Upon the bank of land which breaks the flow
> Of Adria towards Venice:—a bare strand
> Of hillocks, heaped from ever-shifting sand,
> Matted with thistles and amphibious weeds,
> Such as from earth's embrace the salt ooze breeds,
> Is this:—an uninhabitable sea-side
> Which the lone fisher, when his nets are dried,
> Abandons; and no other object breaks
> The waste, but one dwarf tree and some few stakes
> Broken and unrepaired, and the tide makes
> A narrow space of level sand thereon . . .
>
> (1–12)

---

[8] Davie, 142.

[9] See the interesting comparison made between the two passages in Cronin, 114–16.

[10] *The Spirit of the Age: or Contemporary Portraits*, ed. E. D. Mackerness (London and Glasgow, 1969), 271.

An elegant, spare triumph of style, these lines capture the 'flow' of perception, sustaining an even conversational tone that is capable of hinting at undercurrents. The passage celebrates the particularity of time and place; its wording is well focused (the choice and positioning of 'Matted' is an example). But the writing owes its authority as much to the control of caesurae which mirror the hesitations of Julian's thoughts as to the descriptive details of 'salt ooze', 'lone fisher', and 'some few stakes'. The rhythms suggest a meditative condition at odds with yet springing from the discipline of watching the 'ever-shifting' world outside the mind. Holding the merely specific at arm's length, the use of 'Such' in line 6 prepares the reader for its later role in satisfying the poem's need to find emblems. Indeed, the whole landscape is an emblem; even if its meaning is subdued until the reader hears Julian's voice modulate out of the description at line 14, a relish for the sublime has already been conveyed by 'uninhabitable':

> I love all waste
> And solitary places; where we taste
> The pleasure of believing what we see
> Is boundless, as we wish our souls to be . . .
>
> (14–17)

Julian's tone, conversational but exalted, is complex in a way which makes Harold Bloom's account of the landscape as '*not* estranged from the self, and so not seen merely as a portion of the self expelled'[11] seem over-idealized. But that a lover of mankind should delight in the absence of human beings is a quiet rather than insisted irony; if the reader chooses to interpret Julian's love of 'waste | And solitary places' as being an emotion that the poet wishes to qualify, the disarming recognition, 'as we wish our souls to be', seems to cancel the possibility that 'taste' is picked out to expose wish-fulfilment. *The Cenci*, written in a spirit of what at times seems parodic competition with both *Julian and Maddalo* and *Prometheus Unbound*, translates Julian's yearning for infinity into more melodramatic terms: 'I love | The sight of agony' (I. i. 81–2). Arguably, the full-blown parody exposes the latent question mark. Yet if Shelley's tones can weaken the authority of Julian, his intimacy with

---

[11] 'Browning's *Childe Roland*: All Things Deformed and Broken', *The Ringers in the Tower: Studies in Romantic Tradition* (Chicago and London, 1971), 164.

the character is close and tricky, the poetry hovering fruitfully between empathy and judgement.

If the passage after line 14 brings *The Cenci* to mind, ensuing lines travel back to the second stanza of Wordsworth's 'Immortality Ode':

> and yet more
> Than all, with a remembered friend I love
> To ride as then I rode;—for the winds drove
> The living spray along the sunny air
> Into our faces; the blue heavens were bare,
> Stripped to their depths by the awakening North;
> And, from the waves, sound like delight broke forth
> Harmonizing with solitude, and sent
> Into our hearts aërial merriment . . .
>
> (19–27)

'The Moon doth with delight | Look round her when the heavens are bare' (12–13) seems to be remembered by Shelley in line 23, though Julian is, as yet, virtually untroubled by the undertow of loss perceptible in Wordsworth's yearning lines. However, the poetry hints at dissonances of mood, balancing ambitious phrases that suggest the interaction of mind and world against casual rhythms, speaking about a particular experience as if it was typical and repeatable, but catching the experience so specifically that its uniqueness comes to the fore. The gap between the particular and general is significant because it alerts the reader to a struggle which is central to the poem between the raw stuff of experience and the way it is rendered and interpreted. For the most part, this struggle is imaginatively beneficial, directing the reader's attention to nuance and tonal subtlety: even the seemingly nonchalant line, 'The sun was sinking, and the wind also' (35), catches a shade of the mind's self-communion; as the line pauses and quietly broods after the central caesura it achieves a subtler effect than is allowed for by Robert Pinsky's account of Shelley's 'urbane, sensitive, careless spontaneity'.[12] Ultimately the fascination of the writing lies in the fact that it requires the reader to detect a sensibility which cannot be simply ascribed to the character; the poet is less interested in depiction of the limits of character than in momentarily collapsing the distinction between character and poet. The result is not to turn

---

[12] *Landor's Poetry* (Chicago and London, 1968), 15. Pinsky has the Conclusion of *The Sensitive-Plant* most directly in mind.

the characters into mouthpieces of the poet, but to allow the poet freedom to project himself into the imagined experience of the characters.

The argument between Julian and Maddalo is conducted with an easy grace that initially makes one want to ignore the slight irony in Pinsky's generalized description of Shelley's urbanity as 'that of conversation, the natural conversation of aesthetically minded nineteenth-century gentlemen'.[13] But Pinsky's phrasing highlights a crucial issue raised by this poem's use of a conversational idiom: to what extent is Shelley asking us to register the limitations of class conditioning as we listen to the way in which the two friends talk to one another? However, were one to describe the two men as 'aesthetically minded nineteenth-century gentlemen' one would import a cruder class stereotyping into *Julian and Maddalo* than the poem warrants. Shelley is aware of the privileged background of his two characters, but, *pace* Everest, he does not see their thinking as inevitably vitiated by their social position: the tone of his well-known account to Leigh Hunt of his choice of style is realistic rather than snobbish, acknowledging the primary impulse of the poem as being an interest in 'the actual way in which people talk with each other':

I have employed a certain familiar style of language to express the actual way in which people talk with each other whom education and a certain refinement of sentiment have placed above the use of vulgar idioms. I use the word *vulgar* in its most extensive sense; the vulgarity of rank and fashion is as gross in its way as that of Poverty, and its cant terms equally expressive of bare conceptions, and therefore equally unfit for Poetry.[14]

The balance which these sentences attempt to strike between the 'vulgarity of rank and fashion' and 'that of Poverty' does not wholly succeed in extricating Shelley from the web of class. One may regret his inability to recognize the creative possibilities offered by 'vulgar idioms' (as Byron and Clare, in their contrasting ways, were able to do). But Shelley's account has this imaginatively redeeming implication: he will write in an idiom which is rooted in 'the actual way in which people talk with each other'. Not everyone may talk in the way that Julian and Maddalo do, but it is the poem's achievement to represent their ways of talking with subtlety and vigour. His 'familiar style' is flexible enough to accommodate the

---

[13] *Landor's Poetry*, 15.   [14] *Letters*, ii. 108.

characteristic speech of two very different figures. Shelley tries through shifts of tone to ward off the risk that the dialectic may seem flatfooted, and complicates the slightly caricatured stances of idealist and misanthrope:

> Our talk grew somewhat serious, as may be
> Talk interrupted with such raillery
> As mocks itself, because it cannot scorn
> The thoughts it would extinguish . . .
>
> (36–9)

The echo in 'somewhat serious' of the Preface's teasing of Julian as 'rather serious' invites us once more to see the character, Julian, as the focal point for intricate overlaps of tone. Here, Julian's uneasy bravado as he refers to 'thoughts' that lie below and complicate expression is both humanly convincing and representative: *Julian and Maddalo* is not, appearances to the contrary, a poem in which 'thoughts' and 'talk' are on straightforward terms. These lines pass into a passage that wittily but 'forlornly' deploys echoes of the philosophical debate between the fallen angels in Book II of *Paradise Lost* (546–69):

>   'twas forlorn
> Yet pleasing, such as once, so poets tell,
> The devils held within the dales of Hell
> Concerning God, freewill and destiny:
> Of all that earth has been or yet may be,
> All that vain men imagine or believe,
> Or hope can paint or suffering may atchieve,
> We descanted . . .
>
> (39–46)

If the self-conscious elegance of ''twas forlorn | Yet pleasing' strengthens the reader's impulse to read the earlier 'pleasure' (16) critically, it is an impulse that the poetry's own self-awareness forestalls. The feeling that 'descanting' on ideas is an agreeable yet ultimately futile pleasure is entertained by the speaker, but finally dismissed as Shelley retrieves the 'hope' which Milton dismisses as 'Fallacious' (ii. 568). The concession to Milton's stern reference to 'vain wisdom' (ii. 565) in the line about 'vain men' is quickly reversed: the talk concerns as well all that 'hope can paint or suffering may atchieve', and the phrasing shows not only a tension 'between a despairing recognition of human futility, and a celebration of human potential',[15] but also that this tension has been internalized within the speaker. Arguably, this is because, as Wasserman puts it, 'The presence of Maddalo has a disturbing gravitational pull on him.'[16] But Julian's deceptive change of mood—pointed up by the confidential intensity of the parenthetical question that follows: '(for ever still | Is it not wise to make the best of ill?)' (46–7)—makes it intriguingly difficult to reduce *Julian and Maddalo* to a straightforward debate between idealist and pessimist. Indeed, to withdraw into guarded privacy where the reader expects revelation is characteristic of a poem whose moods are at times hard to fathom.

   This unfathomableness gives the poem much-needed depth and suggestiveness. And yet if we want answers, it is the poem which has prompted us to look for them. A vital energy of *Julian and Maddalo* is its search for emblematic correlatives of the human condition. However, 'emblematic' may suggest a cruder poetic technique than Shelley employs. It would be more accurate to say that he dramatizes the longing for certainty in his two main characters. At times the existence of different viewpoints is hinted through the subtle use of transition, as in the movement of these lines from analysis of Maddalo's despair to description of the sunset:

> The sense that he was greater than his kind
> Had struck, methinks, his eagle spirit blind
> By gazing on its own exceeding light.
> —Meanwhile the sun paused ere it should alight,
> Over the horizon of the mountains;—Oh,
> How beautiful is sunset, when the glow
> Of Heaven descends upon a land like thee,
> Thou Paradise of exiles, Italy!
>
> (50–7)

The lines are impressive precisely because of their refusal to point up a clumsily ironic contrast: the use of *rime riche* in lines 52 and 53[17] serves to bring out Julian's undogmatic slide from acute insight into Maddalo's character to awareness of what is ignored by his friend's outlook. Wasserman's discovery of a juxtaposition between 'an ideal light turned blindingly inward upon itself and an ideal light

---

[15] Cronin, 116.   [16] Wasserman, 65.   [17] Pointed out in *PP* 114.

that sheds itself on the world and transfigures it'[18] sacrifices the immediate life of the poetry in favour of a clearer-cut, ultimately constricting thematic account. Competing views of experience lodge themselves within an individual consciousness (whose workings are frankly signalled by the parenthetical 'methinks' and exclamatory 'Oh'), and give the language its vitality. If Shelley's avoidance of simplistic irony is admirable, so, too, is the way he colours Julian's optimistic words and thoughts with darker suggestions: in the last line, for instance, the poetry balances utopian rapture against a recognition that human beings are, indeed, exiled from 'Paradise'.

The description of sunset that follows displays Shelley's ability, vital for the poem's success, to modulate out of a plain style into more heightened evocation; it reveals, too, the way the poetry roots the discovery of value in the workings of consciousness:

>   the hoar
> And aery Alps towards the North appeared
> Through mist, an heaven-sustaining bulwark reared
> Between the East and West; and half the sky
> Was roofed with clouds of rich emblazonry
> Dark purple at the zenith, which still grew
> Down the steep West into a wondrous hue
> Brighter than burning gold, even to the rent
> Where the swift sun yet paused in his descent
> Among the many folded hills . . .
> And then—as if the Earth and Sea had been
> Dissolved into one lake of fire, were seen
> Those mountains towering as from waves of flame
> Around the vaporous sun, from which there came
> The inmost purple spirit of light, and made
> Their very peaks transparent.

>   (67–76, 80–5)

The intricate, melodious couplets organize the vision to which their movement surrenders itself. They are full of unobtrusive hints of Julian's will to convert the scene into an image of the mind's desire. He may try to adapt 'Heaven' (used in this passage at lines 56, 69, and 92) to his secular, humanist perspective, yet the effect is disconcerting. The reader is reminded that Julian's 'heaven' has none of the sureness about its own ontological or epistemological status

---

[18] Wasserman, 70.

of the orthodox 'heaven'. But the language sympathetically implies the strong element of wish-fulfilling hope in Julian's description whose figurative richness, unusual in the poem, discloses the speaker's anxiety to deck out his vision in 'a wondrous hue'. The impulse to imagine transformation is embodied in such details as the dissolving of Earth and Sea 'into one lake of fire'. And yet the possibility of reversing Julian's perspective, so that the sunset is seen less as an image of hope than of transience, is present throughout the passage. This possibility is not managed heavy-handedly by Shelley. For instance, the notion of a 'pause' in Julian's speech (at 53, 63, and 75) induces anxiety that the sunset will pass, but the tranquil simile at lines 63–4 ('As those who pause on some delightful way | Though bent on pleasant pilgrimage') softens menace into something more subliminal. As he does throughout the poem, Shelley establishes a relationship between hope and doubt which is finer than that of contrast. The vision's positive suggestions—the material world both revealing its beauty and dissolving as though spiritualized into 'The inmost purple spirit of light'—are generously imagined.

The poem, then, allows us to experience the tug between conflicting apprehensions. So, a syntactical parallelism describes the effect on Julian of the two contrasting visions (the Venetian sunset and the building brooding against the sky): 'I leaned, and saw the City, and could mark' (89); 'I looked, and saw between us and the sun' (98). The supple suggestion of balance, of Julian's readiness to consider a view antagonistic to his own, supplies another rebuttal to readings of the poem that see it as engaged in adjudicating between ideologies. It may be the case that Julian's irony in lines 111–13 has the callow, even callous ring of one to whom 'maniacs' are as yet abstractions, points of departure for an atheistic gibe rather than suffering individuals: 'As much skill as need to pray | In thanks or hope for their dark lot have they | To their stern maker'. But this view of Julian is not a perception that Shelley chooses to underscore; instead, he brings out Julian's responsiveness to Maddalo's dark vision, a vision whose imaginative energies impress Julian almost as much as distressing him: 'I recall | The sense of what he said, although I mar | The force of his expressions' (130–2). Shelley has, in fact, brought out the 'force of his expressions' by control of tone: the shift from levity to seriousness is managed expertly—'But the gay smile had faded in his eye' (119)—and the vigour with which Maddalo's emblem-hunting imagination works is allowed full scope:

'And such,' he cried, 'is our mortality
And this must be the emblem and the sign
Of what should be eternal and divine!—
And like that black and dreary bell, the soul,
Hung in a heaven-illumined tower, must toll
Our thoughts and our desires to meet below
Round the rent heart and pray—as madmen do
For what? they know not,—till the night of death
As sunset that strange vision, severeth
Our memory from itself, and us from all
We sought and yet were baffled!'

(120–30)

It is true that there is friction between Maddalo's forceful gloom and the qualifying reservations communicated by Shelley's symbols. As 'radiance' (103) is necessary for 'the black bell' (134) to stand out 'In strong and black relief' (106), so when night comes, both 'the black bell' and 'The broad star | Of day' (132–3) are invisible. But Maddalo's vision is neither validated nor ignorable. His lines with their mimetically 'baffled' syntax (127–30) outweigh Julian's anti-Christian contempt; for Maddalo the human predicament is one of perpetual frustration, of irreconcilable clashes between the 'heaven-illumined' soul and the 'rent heart', between 'Our thoughts and our desires' on the one hand, our ultimate ignorance on the other. Shelley succeeds in realizing the force of Maddalo's vision while at the same time suggesting the limitations inseparable from this force; he voices yet qualifies Maddalo's hunger for certainty. This hunger is evident in the phrasing of 'And this must be the emblem and the sign', where 'must be' both sorrows over and insists, and where 'emblem' underlines Maddalo's wish or need to 'read' reality as a system of significant signs. *Julian and Maddalo* is a poem that recognizes the powerful human impulse to construct patterns of understanding and value that offer coherent versions of reality; but it is, too, a work that brings out with vigilant scepticism the provisionality of these patterns. The poem embroils the reader in adjustments of vision, in holding opposed ideas in the mind at the same time without succumbing to the desire to make polemical capital out of the opposition. It accommodates clashes, friction, but it is more concerned with exploration than with offering answers. As Vincent Newey puts it in a fine essay, the poem 'validates, and involves us

in, different ways of experiencing, seeing, and interpreting the world
without allowing total precedence to any'.[19]

Such 'different ways of experiencing' are dramatized with skill and
subtlety in the conversation between the two friends in Maddalo's
palace the following day (141–211). Maddalo's daughter is the next
in the poem's series of emblems. Once more, Shelley arouses different
responses to Julian: he is likeably enthusiastic, praising the girl's eyes
for gleaming 'With such deep meaning, as we never see | But in the
human countenance' (149–50), yet a gap is implied between his
earlier preference for the infinitude of nature and his subsequent
exaltation of the human. Shelley is not undermining Julian. He is
suggesting, even empathizing with, the unstable wobble of views
and feelings that characterize human consciousness and make it
irreducible (even in the case of someone as politically 'committed'
as Julian) to one overriding idea. That said, Julian does speak with
an ardour that we register as theoretical; it is a style of speaking that
Shelley captures well:

> 'Mine is another faith'—thus much I spoke
> And noting he replied not, added: 'See
> This lovely child, blithe, innocent and free;
> She spends a happy time with little care
> While we to such sick thoughts subjected are
> As came on you last night—it is our will
> That thus enchains us to permitted ill—
> We might be otherwise—we might be all
> We dream of happy, high, majestical.
> Where is the love, beauty and truth we seek
> But in our mind? and if we were not weak
> Should we be less in deed than in desire?'
>
>   (165–76)

The fervent tone and wording of this passage prompt the reader's
scepticism, and relief is afforded by the crisp no-nonsense of Maddalo's
rejoinder: 'You talk Utopia' (179). Yet Richard Cronin is too hasty,
in my view, in wanting to denounce Julian's opening gambit: 'In being
used as an illustration the child degenerates from a human being to
a stereotype.'[20] Unquestionably, the child is 'being used as an
illustration', but it is central to Shelley's undogmatizing sympathy

---

[19] 'The Shelleyan Psycho-Drama: "Julian and Maddalo"', *E OS* 84.
[20] Cronin, 120.

with his characters that he recognizes the virtual inevitability of using experience to illustrate preconceptions about experience. Maddalo's mixture of affection and exasperation—' "My dear friend," | Said Maddalo, "my judgement will not bend | To your opinion, though I think you might | Make such a system refutation-tight | As far as words go" ' (191–5)—cannot but shape our response to Julian here. This is not to exonerate Julian from the blindnesses involved in seeing experience as illustrating the truth of his idealistic philosophy, nor is it to set up Maddalo as the more trustworthy guide to Shelley's view of existence. Instead, it is to insist that the poem declines the hotly eager processes of 'unmasking' and 'subverting' that some critics detect. *Julian and Maddalo* is more concerned to dramatize the clash between theory and experience, between reality and what language makes of it.

It is this dual clash which comes into fascinating prominence in the central section of the poem, the Maniac's monologue overheard by the two friends who, at Maddalo's instigation, visit him. Indeterminacy and frustration are the dominant and calculated impressions made by this section, sometimes criticized as being indulgent and vague. Such criticism is insensitive to the poetry's risk-taking empathy with a difficult, confused emotional state, a state which puts Shelley's conversational style under pressure. In the very way he phrases the following critique, Hazlitt inadvertently suggests Shelley's ability to evoke emotional blocks, 'interferences': 'The depth and tenderness of his feelings seems often to have interfered with the expression of them, as the sight becomes blind with tears.'[21] Certainly, the poem would be more lucid, decorous, and comfortable without the Maniac's outpourings, but it would also be less challenging. When Maddalo formulates this definition of poetry, 'Most wretched men | Are cradled into poetry by wrong, | They learn in suffering what they teach in song' (544–6), the language, as it tries to master suffering through an aphoristic turn of phrase, bears witness to the poetry's sense of the troubling gap as well as link between 'suffering' and 'song'.

There are two ways in which this section is indeterminate: the first is the relationship it posits, or rather declines to posit, between the Maniac's experience and the opposing views of Julian and Maddalo.

---

[21] Review of *Posthumous Poems*, in Redpath, *The Young Romantics and Critical Opinion*, 392.

Before the encounter the Maniac's office of exemplum is quite clear. Maddalo sees the Maniac's fate as proof of idealism's folly—'his wild talk will show | How vain are such aspiring theories' (200–1)—where the generalizing use to which 'such' is put throughout the poem is once more evident. Julian is equally guilty of 'using painful human experience as a dumbbell with which to exercise the moral imagination';[22] to him the Maniac sounds like a case of insufficient will-power and resolve, one that provokes sympathy but leaves the observer in no doubt that distress is the result of 'man's own wilful ill' (211). Undoubtedly both characters are over-glibly ready to explain, and there is a certain felicity in the fact that Julian's speech finishes with the only unrhymed line (211) in *Julian and Maddalo* as the poem leaves theory abruptly behind. After they have heard the Maniac, however, the two friends sink their differences—'our argument was quite forgot' (520)—in a common pity.

Agreeing that his was 'some dreadful ill' (525) but 'unspeakable' (526), Julian and Maddalo set up the extreme terms by which experience and language seek to validate but are shown as frustrating one another in this section of the poem. Shelley probably had this controversial section in mind when he outlined the proper place of the 'familiar style' to Hunt: 'Not that the familiar style is to be admitted in the treatment of a subject wholly ideal, or in that part of any subject which relates to common life, where the passion exceeding a certain limit touches the boundaries of that which is ideal.'[23] A 'passion exceeding a certain limit', the Maniac's self-laceration is recorded with an empathy which shades into exposure of emotional confusion. This shading or overlapping is responsible for the second area of indeterminacy that surrounds the presentation of the Maniac. The poetry shifts between investigating a tortuous as well as tortured mind and suggesting that the issues are crystal-clear. The verdict arrived at by Julian and Maddalo illustrates this intermittent, simplifying clarity: they decide that the Maniac had been wronged by 'a dear friend' (527) 'For whose sake he, it seemed, had fixed a blot | Of falshood on his mind which flourished not | But in the light of all-beholding truth' (529–31). Here, only the hesitation in 'it seemed' suggests that the mechanism of fixing a blot of falsehood on his mind may misrepresent what happened between the Maniac and his 'dear friend'. It is at this point that Maddalo's aphorism about

---

[22] Cronin, 120.   [23] *Letters*, ii. 108.

'suffering' and 'song' (544–6) masks what elsewhere in the poem has been experienced as a more complicated relationship between language and feeling. But Shelley does not criticize Julian and Maddalo for choosing to idealize the Maniac by seeing him as a victim of his own tenderness of feeling, susceptibility to pain. It suits both Julian's belief in the all-important power of the human mind and Maddalo's conviction of the tragic disparity between mind and circumstance to agree that the Maniac has been undone by a sensitivity inseparable from a nobility of spirit that still makes itself felt: 'The colours of his mind seemed yet unworn' (540).

However, the Maniac's speech is a bewildering and powerful tissue of contradictions. If it is confessional, it is also full of obliquities, as in these lines which acknowledge that the speaker is engaged in composition yet declare words to be 'vain':

> How vain
> Are words! I thought never to speak again,
> Not even in secret,—not to my own heart—
> But from my lips the unwilling accents start
> And from my pen the words flow as I write,
> Dazzling my eyes with scalding tears . . . my sight
> Is dim to see that charactered in vain
> On this unfeeling leaf which burns the brain
> And eats into it . . . blotting all things fair
> And wise and good which time had written there.
>
> (472–81)

The violence of this is both explained and justified by the speaker's unwillingness to betray feelings through grappling with words. Context, too, is important; far from being an embarrassment in an otherwise controlled and lucid poem, the Maniac's speech brings out the difficulty of and necessity for a way of speaking that attempts to lay bare the self-communing heart. The Maniac's plight provokes both fear and recognition: fear in that it represents a state of self-enclosed suffering, recognition in that it makes open the fact (the scandal) of vulnerable subjectivity which has been hinted at elsewhere in the poem (most notably during Maddalo's vision of human entrapment). Shelley portrays a mind that oscillates between self-justification and guilt, anger and forgiveness, the impulse to declare and the pressure to conceal. These tugs of feeling are present in the passage after line 438, where the Maniac speaks of the cruelty of

making 'love the fuel | Of the mind's hell' (440–1). Part of that 'hell'
for the speaker is his sense of peculiar susceptibility to suffering; in
lines that are self-dramatizing yet poignant, he refers to himself as
'a nerve o'er which do creep | The else unfelt oppressions of this earth'
(449–50).

There are failings and falsities of tone. Over-emotiveness, shown
by the frequent use of 'and' (275–91) as well as by phrases such as
'his lean limbs shook' (279), disfigures the introduction of the
Maniac. But, elsewhere, it would be wrong to blame Shelley for
presenting a mind anxious to vindicate itself; his job, done with
persuasive attention to the stops and starts of feeling, is to render
the Maniac's emotional states. Compulsive rationalizing gives the
writing its distinctive tone. The following lines, for example—for
all their allusion to a Shakespearean sonnet (CXI) in lines 350–1—
do not support Newey's discovery of some 'heightened, almost
Shakespearean, conception of the workings of scorn and hate':[24]

> Yet think not though subdued—and I may well
> Say that I am subdued—that the full Hell
> Within me would infect the untainted breast
> Of sacred nature with its own unrest;
> As some perverted beings think to find
> In scorn or hate a medicine for the mind
> Which scorn or hate have wounded—o how vain!
> The dagger heals not but may rend again. . . .
>
> (350–7)

What distinguishes the Maniac from the speaker of Shakespeare's
sonnet is the full-blown nature of his self-justifying. In the sonnet
the lines on which the Romantic poet draws—'And almost thence
my nature is subdu'd | To what it works in, like the dyer's hand'
(6–7)—are ruefully self-accepting and figuratively vital, as Shelley
recognized in his comment on them: 'Observe these images, how
simple they are, and yet animated with what intense poetry and
passion.'[25] There is intense passion in Shelley's lines, yet it has little
to do with the workings of imagery. In the Maniac Shelley dramatizes
the gap between emotional pain and the impotence of consciousness
to do anything other than supply theoretical glosses. Such a gap both
underpins and undermines the claim that follows these lines: 'I am

[24] 'The Shelleyan Psycho-Drama', 89.   [25] Julian, vii. 152.

ever still the same | In creed as in resolve' (358–9), as well as
the distinction between 'heart' and 'understanding' on which the
claim depends: 'what may tame | My heart, must leave the under-
standing free' (359–60). Both sympathy and scepticism are incited
by the Maniac's attempts to sustain a stoical front, as in the empty
distinction he makes between his indifference to his own 'repose' and
his inability to 'bear more altered faces':

> not for my own repose—
> Alas, no scorn or pain or hate could be
> So heavy as that falshood is to me—
> But that I cannot bear more altered faces
> Than needs must be . . .

> (309–13)

It is the dramatic accuracy with which Shelley captures the
Maniac's frustrated struggle to reach the centre of his being that keeps
one reading, the pain involved in the speaker's prolonged act of
tearing a veil from his pent mind. The tussle between exposure and
concealment that this leads to is what compels interest rather than
the content of the Maniac's confession (though there is enough
particularizing detail to hold our attention: the passage in which the
speaker records the sexual revulsion experienced by his friend, lines
420–38, supplies a vivid example). Shelley is not being coy or
tantalizing in emphasizing the difficulty of articulating feeling.
Throughout, the poem has concerned itself with the unavoidable
necessity of interpreting and the impossibility of doing so undistort-
ingly. The Maniac's concluding speech achieves neurotic intensity
through obsessive images of concealment and covering, a concern
of the brief scene that Shelley wrote for his proposed drama on
Tasso: '*His grace* | Is buried in deep converse with the dead'.[26] In
opposition to the final movement of the 'Ode to the West Wind',
the Maniac cries, 'I do but hide | Under these words like embers,
every spark | Of that which has consumed me' (503–5), a cry in
which the need for language and sense of its inadequacy fuse. His
final words, 'the air | Closes upon my accents, as despair | Upon
my heart—let death upon despair!' (508–10), win a despairing
eloquence from the histrionic. The section's inconclusiveness, like

---

[26] G. M. Matthews, 'A New Text of Shelley's Scene for *Tasso*', *KSMB* xi
(1960), 40.

that of the entire poem, bears witness to Shelley's discovery—a discovery embodied with great resource in the poem's language and structure—of the relativist, dramatic, irreconcilable nature of experience.

Comparison with Byron's *The Lament of Tasso* (read by Shelley in 1817 and undoubtedly an influence on his poem) helps to point up the achievement of this section of *Julian and Maddalo*. The parallels between Byron's poem and the Maniac's speech are evident, and form the basis of Carlos Baker's argument that 'the key poem in solving the mystery of the identity of Shelley's maniac is Byron's *Lament of Tasso*'.[27] But Shelley chooses to leave the identity of the Maniac in doubt, presenting his character as a fiction, by contrast with Byron who alludes to supposed fact. Generally, where Byron is clear-cut, Shelley is elusive. Byron's Tasso is as declamatory and self-justifying as the Maniac, but where Shelley explores derangement, Byron allows his speaker to define his state of mind with robust common sense: 'my frenzy was not of the mind' (52). Shelley strains beyond the limits of his idiom as Byron never does in *The Lament of Tasso*. Both poets share an obsession with what in *The Corsair* Byron calls 'That opening sepulchre—the naked heart' (ii. 355), but of the two it is Shelley who proves the more exploratory, offering in the Maniac a complex attempt to examine a repressed consciousness.

The conclusion of *Julian and Maddalo* sustains the inconclusiveness that I have been representing as a virtue of the poem. Again, we are encouraged to search beneath Julian's low-key phrasing for a deeper exploration of his motives. Yet we are frustrated in our desire to arrive at an unambiguous view of these motives, the kind of view that would, for instance, make possible condemnation of Julian for uttering the following lines (in which compassion seems to be distanced as one of the poem's emblems):

> and this was all
> Accomplished not; such dreams of baseless good
> Oft come and go in crowds or solitude
> And leave no trace—but what I now designed
> Made for long years impression on my mind.
>
> (577–81)

---

[27] Baker, 135.

By contrasting the 'impression' his plan left on his mind with 'such
dreams', Julian tries to assert his sincerity, yet his distinction seems
only to point up a lack of difference. The reader is free to catch at
a whisper of criticism in the lines, 'and this was all | Accomplished
not', but is, nevertheless, conscious of wanting to resolve issues that
Shelley chooses to leave unsettlingly in the air. Throughout *Julian
and Maddalo*, Shelley has skilfully dramatized the penumbra of
possibilities that surrounds moods, judgements, and actions; here,
Julian describes his encounter with Maddalo's daughter in a way that
suggests growing disillusion and residual idealism:

> His child had now become
> A woman; such as it has been my doom
> To meet with few, a wonder of this earth,
> Where there is little of transcendent worth,
> Like one of Shakespeare's women . . .

(588–92)

The use, once again, of 'such' alerts the reader to the inescapable
human drive to interpret; the detail does not ask us to be sceptical
about the 'transcendent worth' of Maddalo's daughter, but merely
requires us to acknowledge the inevitable and never-ending quest
to find some accord between ideals and experience. It is an open-
ended effect; yet if the entire ending is open in one sense (nothing
is resolved), it is closed in another (nothing is revealed). Maddalo's
daughter could serve as evidence either that her father's pessimism
self-thwartingly ignored 'transcendent worth' or as pointing up the
sad rarity of human excellence. But in fact the argument has long
since glimpsed its ultimate unresolvableness. The comparison with
'one of Shakespeare's women' is both a shining compliment and a
tacit admission that such perfection is more commonly glimpsed in
art than in life. To see Maddalo's daughter as kindling hope in the
midst of 'meaningless and prosaic indifference'[28] is an interpretative
possibility that the poetry neither dismisses nor wholly endorses.
What can be stated unequivocally is that the suspended, withheld
quality of the conclusion is impressively communicated through
Shelley's language, a considerable achievement from a writer often
thought of as hurrying towards certainties. It is fitting that the poem's
ending (where Maddalo's daughter finally tells Julian about the

---

[28] Wasserman, 81.

Maniac's fate) should offer another version of its pattern of disclosure and reticence. The lines send out further ripples of suggestion, hinting at some ultimate revelation, yet positioning the reader between the not fully comprehensible speaker (Julian) and the uncomprehending 'cold world':

> I urged and questioned still, she told me how
> All happened—but the cold world shall not know.

(616–17)

4

The Cenci
Language and the Suspected Self

The Cenci and *Julian and Maddalo* are both works which combine
an exploration of subjectivity with a capacity to 'construct', as
Vincent Newey puts it, 'an effect of objectivity'.[1] Yet in neither
work does this 'objectivity' take the form of explicit authorial
judgement. Indeed, there is a sharp gap between Shelley's description
of *The Cenci* as portraying 'a sad reality' (*PP* 237) and the system
of ideals that many commentators suppose the play to be invoking,
if only as a significant absence. For instance, Michael Scrivener
contends that 'The spectator of the tragedy realizes that only a
radically different society, based on principles that transcend
patriarchy, could permit the emergence of an innocence which
does not contradict itself.'[2] Such a realization might seem to be
implicitly sponsored by Shelley's own analysis in the play's Preface
of his heroine's conduct: 'Revenge, retaliation, atonement, are
pernicious mistakes. If Beatrice had thought in this manner she
would have been wiser and better'. There, however, the ethical
treatise-writer breaks off and the imaginative poet takes over:
' . . . but she would never have been a tragic character' (*PP* 240).
The 'tragic' for Shelley involves an exploration of contradictions;
in *The Cenci* such exploration usurps any desire to 'familiarise the
. . . imagination of . . . poetical readers with beautiful idealisms of
moral excellence' (*PP* 135). In *A Defence of Poetry* he explains the
appeal of tragedy in this way: 'Sorrow, terror, anguish, despair itself
are often the chosen expressions of an approximation to the highest
good. Our sympathy in tragic fiction depends on this principle;
tragedy delights by affording a shadow of the pleasure which exists
in pain.' (*PP* 501.)

---

[1] 'The Shelleyan Psycho-Drama', 96.
[2] *Radical Shelley: The Philosophical Anarchism and Utopian Thought of Percy
Bysshe Shelley* (Princeton, NJ, 1982), 196.

Shelley is made uneasy by and tries to offer a justification for the aesthetic pleasure given by 'tragic fiction'. But both the ethical commentator of the Preface and the aesthetic theoretician of *A Defence* do not so much explain as explain away the specific power of *The Cenci*. The play suspends belief in absolutes and explores what life seems like to the imagination once the mind is impelled to test its hopes, desires, ideals. Indeed, one ideal which the play enmeshes in contradictions is the notion that the mind can fathom itself, can arrive at illumination and truth. In *A Defence* Shelley links scepticism about the mind's capacity for unmediated intuition of its own workings ('Neither the eye nor the mind can see itself, unless reflected upon that which it resembles') to a celebration of 'drama' as 'a prismatic and many-sided mirror' (*PP* 491). Such a mirror reflects, so Shelley's argument runs, the mind that brings it into being. In a recent essay Stuart Curran attempts to retrieve the play from complete existential absurdity by invoking what he calls Shelley's 'unitary faith', which he sees as working to warn us to 'Beware of converting fictions into fixities. Beware—and be free.'[3] By contrast, this chapter will contend that *The Cenci* is a mirror in which the mind beholds contradictions and that the leap from fixities to freedom is not taken by the play. Certainly, Shelley unmasks the workings of patriarchal tyranny with a deadly clarity—on occasions, too deadly: there are times when the play verges on caricature in its presentation of Cenci. But though *The Cenci* exposes error and injustice, it is (to its artistic advantage) far less ready to offer unambiguous examples of or signposts towards truth and goodness. Indeed, it is Shelley's readiness to explore and remain in the midst of doubt and uncertainty, to be, in short, more negatively capable than is usually allowed, that is impressive. Orsino's falsely assured claim—'I see, as from a tower, the end of all' (II. ii. 147)—is only the most conspicuous example of irony directed against those who believe they can see 'the end of all'.

In both *Julian and Maddalo* and *The Cenci*, Shelley's own linguistic concerns could be said to be projected into the plot. These concerns are twofold in nature: the obsession with gaps between thought and speech, and the attempt to find an imaginative idiom capable of communicating this obsession. Both concerns are traceable in the

---

[3] 'Shelleyan Drama', in Richard Allen Cave (ed.), *The Romantic Theatre: An International Symposium* (Gerrards Cross, Bucks., and Totowa, NJ, 1986), 76, 77.

poems of 1816, but both take on darker intensity in *Julian and Maddalo* and *The Cenci*. The two works are troubled by the 'dark estate' (*Julian and Maddalo*, 574) of the human condition. In each the central event is too painful to be articulated directly. The Maniac has suffered some 'dreadful ill' (525) which was 'yet unspeakable' (526); Beatrice has 'endured a wrong, | Which, though it be expressionless, is such | As asks atonement' (III. i. 213–15). As Michael Worton points out, 'there is a consistent and deliberate refusal by the playwright to *name* the catalysing action within the tragedy'.[4] Shelley's refusal to name this 'catalysing action' can seem over-obvious, as when Lucretia says of Orsino's note: 'It speaks of that strange horror | Which never yet found utterance' (IV. iv. 97–8). Yet, if this kind of writing rouses the very voyeurism which Shelley, desiring to 'increase the ideal, and diminish the actual horror of the events' (*PP* 239), wishes to hold at bay, it is unquestionably the case that the dramatist and his characters are united by a common awareness of the difficulty of 'utterance'.

Shelley wishes to arouse an impulse of 'restless and anatomizing casuistry' (*PP* 240) in the reader; whether such 'casuistry' acts as the springboard for enlightenment or turns into a maze which entraps is a major issue raised by the play. Though Worton accepts that 'the poet . . . offers no solution to the moral dilemma of Beatrice's guilt or innocence',[5] he claims that the play 'seeks to engender self-awareness on the part of the audience by the radically modern device of calling itself into question'.[6] Yet it is more accurate to regard the play as deconstructing the idea of 'self-awareness'; by exposing the dependence of awareness on language, with all its pitfalls and treacheries, the play confronts the audience with the impossibility of arriving at a stable sense of self. The drama enacts Shelley's pessimistic assertion, 'Words are the instruments of mind . . . but they are not mind, nor are they portions of mind',[7] and indeed the pitiless unmasking of certainties about self and language explains *The Cenci*'s distinction.

While *The Cenci* is, in Bloom's phrase, 'a work conceived in the Shakespearean shadow',[8] it is not the case that 'Shelley did not

[4] 'Speech and Silence in *The Cenci*', *EOS* 107.
[5] 'Speech and Silence in *The Cenci*', 120.
[6] 'Speech and Silence in *The Cenci*', 120.
[7] 'Speculations on Metaphysics', Julian, vii. 63.
[8] 'The Unpastured Sea: An Introduction to Shelley', *The Ringers in the Tower*, 105.

succeed in forming a dramatic language for himself in his play'.[9]
The first lines of the play, Camillo's wheedling, secretive bargaining
with Cenci, display a stylistic sureness which runs through the work:

> That matter of the murder is hushed up
> If you consent to yield his Holiness
> Your fief that lies beyond the Pincian gate.

> (I. i. 1–3)

Both the bareness and readiness to trust in implication of this passage
are striking and successful. The memorable first line wins from its
alliterative meshing of 'matter' and 'murder' an effective suggestion
of complicity's ineffectual muttering. Although the Shakespearean
echoes in the play are unignorable, they are often used intelligently
to point up thematic similarities and contrasts. Discussing the 1959
Old Vic production of the play, Richard Allen Cave discerns in
Barbara Jefford's interpretation of Beatrice a view of the character
which is close to mine: 'She [Beatrice] made a bid for freedom, but,
rather like Macbeth, destroyed all pattern and meaning in the
universe as a consequence. (This is one interesting way of handling
the numerous "echoes" of *Macbeth* throughout the text of *The Cenci*,
making them an ironic commentary on the fate of the heroine.)'[10]
This way of handling the echoes is not only 'interesting' but also
properly respectful of Shelley's allusive intelligence, amply
demonstrated throughout his career. None the less, it is true that
such intelligence does not prevent the verse from being occasionally
swamped by the Shakespearean influence.[11]

But Shelley forms a dramatic language for himself with
considerable success whenever he focuses, self-consciously, on
language. In Act V, for instance, Giacomo expresses the perverse
way language can batten on to and control thought:

> O, had I never
> Found in thy smooth and ready countenance
> The mirror of my darkest thoughts; hadst thou
> Never with hints and questions made me look

[9] 'The Unpastured Sea: An Introduction to Shelley', 105.
[10] 'Romantic Drama in Performance', *The Romantic Theatre: An International Symposium*, 92.
[11] Leavis's analysis of Giacomo's soliloquy in III. ii., for example, is rightly critical both of its dependence on *Othello* and of its 'insistent externality', Leavis, 222.

Upon the monster of my thought, until
It grew familiar to desire . . .

(V. i. 19–24)

The mind's accommodation to what it feels to be evil finds apt
expression in the subtle, insinuating syntax. Giving a grimly parodic
twist to Shelley's assertion in *Prometheus Unbound* that 'speech
created thought' (II. iv. 72), these lines represent language as fatally
domesticating the alien and as alienating the self from its truest
impulses. The wording may owe a debt to a passage in Coleridge's
Remorse, where Valdez says to Ordonio: 'O that I ne'er had yielded
| To your entreaties!' (III. ii. 88–9);[12] elsewhere in his play Coleridge
shows an interest in the relationship between thought and deed:
'speaking of himself in the third person',[13] Ordonio asserts: 'Some-
thing within would still be shadowing out | All possibilities' (IV. i.
113–14). But what makes Shelley's lines superior to both passages
in *Remorse* is the way his phrasing is alive to the possibility
that pre-linguistic thoughts can be unearthed, and their potential
frighteningly released, by 'hints and questions'. It is, as Wasserman
observes, a view that runs 'in exact contradiction to all theories that
psychoanalysis is cathartic'.[14]

Giacomo's speech offers a particular instance of Shelley's more
general account in the Preface of language in drama: 'In a dramatic
composition the imagery and the passion should interpenetrate one
another, the former being reserved simply for the full development
and illustration of the latter. Imagination is as the Immortal God
which should assume flesh for the redemption of mortal passion.'
(*PP* 241.) In the above example, 'imagery', bare and functional,
reinforces 'passion'. The image of the mirror is used to imply the
treacherous accord between barely existing thoughts and another's
promptings. In his curse earlier in the play (IV. i.), Cenci prays that
Beatrice's child might be a 'distorting mirror' (147) wherein she will
find 'A hideous likeness of herself' (146), invoking a distortion
equivalent to that which Giacomo sees as being inflicted on thought
by language. The force of the implicit analogue suggests the play's
conceptual energy.

---

[12] Quoted from vol. ii of *The Complete Poetical Works of Samuel Taylor
Coleridge*, ed. E. H. Coleridge (2 vols., Oxford, 1912).
[13] *The Complete Poetical Works of Samuel Taylor Coleridge*, ii. 862.
[14] Wasserman, 111.

This conceptual energy is intimately linked to the writing's awareness of the limits of language. Mental states are described in a style
whose generalizing confidence is belied by anxiously scrupulous
qualifications:

> and thus unprofitably
> I clasp the phantom of unfelt delights
> Till weak imagination half possesses
> The self-created shadow.
>
> (II. ii. 140–3)

Here, the very insubstantiality of Orsino's generalizations captures
his state of unfulfilled wish-fulfilment. A less successful example of
psychological notation in the play is offered by the same character's
words in reply to Giacomo's speech already quoted from Act V:

> confess 'tis fear disguised
> From its own shame that takes the mantle now
> Of thin remorse.
>
> (V. i. 30–2)

Orsino seeks to explain away Giacomo's words as cowardly self-
deception, and yet the subtlety is not without a self-thwarting
element as the language invites us to imagine a 'disguised' abstraction
assuming the guise of another abstraction.

Both Shelley's fascination with inner states of mind and his sense
of the restrictions of language are evident in the Preface where
he refers ironically to the Protestant 'passion for penetrating the
impenetrable mysteries of our being, which terrifies its possessor at
the darkness of the abyss to the brink of which it has conducted him'
(*PP* 240). The sonorous phrasing acquiesces in linguistic frustration.
Again, Shelley writes confidently about his enterprise, speaking of
it as 'a light to make apparent some of the most dark and secret
caverns of the human heart' (*PP* 239); the play itself, however, is
most illuminating when most aware of the impenetrability of the
mysteries of our being, an awareness which gives rise to a continual
tension in the language. When, for instance, Orsino refers to 'the
inmost cave of our own mind' (II. ii. 89) he is trying to hoodwink
Giacomo, yet his wording pinpoints a quarrel between the irreducibly
private ('inmost') and the universally shared ('our own mind') which
is central to *The Cenci*'s presentation of states of mind. For all his
insistence that the language used in poetry 'must be the real language

of men in general and not that of any particular class to whose society the writer happens to belong' (*PP* 241–2), Shelley relies at times on stylized rhetoric as a way of universalizing his play's themes, as in these finely cadenced lines in II. ii: 'And we are left, as scorpions ringed with fire, | What should we do but strike ourselves to death?' (70–1). However, the power of Giacomo's lines (spoken in the course of dialogue with Orsino) has to do with the character's escape from inner disturbance into arresting self-dramatization: his images of circling fire and self-destroying scorpion evoke a nightmarish sense of menace and entrapment.[15] Shelley uses rhetoric purposefully here both to win an audience's sympathy for the helplessness felt by Giacomo and to alert it to the character's quickness to give up hope.

But what follows (Orsino's attempt to coax Giacomo into speech) presses close to the heart of the play's concern with language:

> Fear not to speak your thought.
> Words are but holy as the deeds they cover:
> A priest who has forsworn the God he serves;
> A judge who makes truth weep at his decree;
> A friend who should weave counsel, as I now,
> But as the mantle of some selfish guile;
> A father who is all a tyrant seems,
> Were the prophaner for his sacred name.
>
> (II. ii. 74–81)

This speech suggests the unreliability of language, and belies in the use of 'cover' the identity of word and deed which it seems to imply. Orsino's distrust of 'words' is an artful fraud (his own successfully fool Giacomo) that ironically states a central doubt of the play. In these lines and in Giacomo's reply, Shelley finds a style in which to explore the characters' feelings and anxieties. Giacomo's reply shows *The Cenci*'s capacity to guide us towards the operations of the unconscious while respecting their otherness:

> Ask me not what I think; the unwilling brain
> Feigns often what it would not; and we trust
> Imagination with such phantasies
> As the tongue dares not fashion into words,

---

[15] See Keach, 106.

Which have no words, their horror makes them dim
To the mind's eye.

(82–7)

Here, as elsewhere in *The Cenci*, the way language works mirrors
the play's thematizing of the workings of language. Giacomo's
concern not to 'fashion into words' the brain's 'phantasies' runs
parallel to Shelley's concern to express the character's sense of the
need to keep his deepest promptings unexpressed. The effect is to
substitute slow-motion brooding for action, and the passage may
risk being excessively self-conscious. Again, however, Shelley's
language satisfyingly conveys Giacomo's state of mind; his mental
torture is articulated by a syntax that enacts the eruption and
repression of 'phantasies'. Cenci anticipates the mental processes
explored here, when in the opening scene he confesses gloatingly
to Camillo how 'This mood has grown upon me, until now | Any
design my captious fancy makes | The picture of its wish' (86–8).
However, Cenci feels none of Giacomo's self-revulsion; wishes
picture themselves with a glib ease which suits his simplified if
unnerving characterization. The writing compares favourably with
Byron's use in *Marino Faliero* of a more elaborated version of the
same idiom:

> though as yet 'tis but a chaos
> Of darkly brooding thoughts: my fancy is
> In her first work, more nearly to the light
> Holding the sleeping images of things
> For the selection of the pausing judgement.—

(I. ii. 282–6)

'Darkly brooding thoughts' is the kind of stereotyped phrase
that Shelley sensibly avoids, though in the case of Cenci he is able
to suggest the presence of such thoughts. A character whose life,
in Stuart Curran's words, 'is perpetrated on the utter negation of
life'[16] represents an extreme that poses a challenge to psychologiz-
ing ambitions. Does Cenci embody a force of evil that resists
explanation? For the most part Shelley creates a chilling balance
between suggesting that the roots of Cenci's behaviour are ultimately
inexplicable and dramatizing the will to power over others which

---

[16] *Shelley's 'Cenci': Scorpions Ringed with Fire* (Princeton, NJ, 1970), 86.

has become a need for the Count. The speech where he asserts his
will over Beatrice is a frightening and achieved example:

> Stay, I command you—from this day and hour
> Never again, I think, with fearless eye,
> And brow superior, and unaltered cheek,
> And that lip made for tenderness or scorn,
> Shalt thou strike dumb the meanest of mankind;
> Me least of all.

(II. i. 115–20)

The swagger of 'I think' and the suppression of injured pride in
the final phrase make the speech a convincing piece of emotional
bullying. The difficulty for Shelley is to sustain both the domestic
truth and the wider implications of Cenci's conduct and vision. At
times these wider implications are conveyed over-explicitly to an
audience, as when Cenci is made to invoke 'God, the father of all'
(I. iii. 118). But there is enough power and insight in Shelley's
presentation of Cenci to make Stuart Curran's hyperbole ('stunning
in its perception')[17] seem not unfounded.

Indeed, two compelling moments in the play—the 'dread abyss'
(III. i. 254) which Beatrice describes and 'the pit' (II. ii. 114) which
Orsino says Cenci fell into—owe their force to the play's exploratory
sense of evil. Orsino's speech which contains this reference to Cenci
is both coolly self-incriminating and sensitive to the gap between
psyche and language:

> . . . 'tis a trick of this same family
> To analyse their own and other minds.
> Such self-anatomy shall teach the will
> Dangerous secrets: for it tempts our powers,
> Knowing what must be thought, and may be done,
> Into the depth of darkest purposes:
> So Cenci fell into the pit; even I,
> Since Beatrice unveiled me to myself,
> And made me shrink from what I cannot shun,
> Shew a poor figure to my own esteem,
> To which I grow half reconciled.

(II. ii. 108–18)

---

[17] *Shelley's 'Cenci': Scorpions Ringed with Fire*, 86.

Orsino incriminates himself in the depressed cadences of the concluding lines, and he incriminates others in the 'must' of 'Knowing what
must be thought'. But the speech is the more intense for its lack of
overt emotional intensity. There is little horror in Orsino's glimpse
of 'darkest purposes', the phrase representing the merest gesture
towards the barely formulable. Paradoxically, the effect is not to
diminish but to increase the impact of the speech, which catches in
its tones and diction the destructively objective view of self that
Orsino is saying is the risk courted by 'self-anatomy'. It is true, as
Stuart Curran argues, that 'Evil multiplies in *The Cenci* with such
intensity that it seems self-generating.'[18] Yet 'evil' is most credible
in the play when least melodramatic, as in these lines, with their
trenchant account of the 'Dangerous secrets' that lie in store for the
over-introspective.

Orsino's language is animated by the play's fascination with mental
processes and, more particularly, the ease with which they can be
perverted. The Preface sets up 'teaching the human heart . . . the
knowledge of itself' (*PP* 240) as a bulwark against and alternative
to 'self-anatomy', but the play itself calls into question the very idea
of the self. Shelley adapted lines from Byron's *Manfred*—'But I can
act even what I most abhor, | And champion human fears' (II. ii.
203–4)—which haunt Beatrice's final degree of complicity in evil.
In Shelley's variation on Byron's lines Cenci is made to say:

> She shall become (for what she most abhors
> Shall have a fascination to entrap
> Her loathing will), to her own conscious self
> All she appears to others . . .
>
> (IV. i. 85–8)

The quickening movement of 'fascination' allows the reader to
experience from within the convolutions which, elsewhere, Shelley
includes under the heading of 'self-contempt'.[19] A desire to believe
in the determining power of the will often asserts itself in Shelley's
work: 'it is our will | That thus enchains us to permitted ill' (*Julian
and Maddalo*, 170–1). Here Cenci presents a darker, frightening
vision of the mind as a place where 'loathing will' competes with

---

[18] *Shelley's Annus Mirabilis: The Maturing of an Epic Vision* (San Marino, Calif.,
1975), 133.
[19] See Wasserman for a lucid discussion of 'self-contempt', pp. 109–10.

unconscious 'fascination'. Certainly Beatrice relies on what earlier in the scene Cenci calls 'her stubborn will' (10)—not, however, as Cenci intends, to consent to violation, but to resist it, even if it is arguable that she is shown as corrupting herself in the act of opposing incestuous rape with parricide. Yet, though the play is fascinated by characters who abuse will-power and discover the limits of the will, it is too simplistic to see *The Cenci* as ultimately showing that the will, rightly directed, results in good. Shelley's drama is dominated by the fear that the self, far from being inviolable, is in fact highly vulnerable and resists being thought of predominantly in terms of the 'will'.

The audience is persuaded to sense that Beatrice's 'self' is lost to her father as he sadistically relishes his mastery of it. Indeed, a self more private than the self as it appears to others is experienced both as presence and absence in *The Cenci*. So in a later scene Beatrice says to Lucretia:

> 'Tis like a truant child
> To fear that others know what thou hast done,
> Even from thine own strong consciousness . . .

>   (IV. iv. 36–8)

'Consciousness' may be the source of fear that others can read us like a book, but it is also, Beatrice's words imply, a possible refuge from others.

The Cenci contains fascinating examples of self-conscious fascination with consciousness. Beatrice's words to Lucretia in III. i. offer an illustration:

> What are the words which you would have me speak?
> I, who can feign no image in my mind
> Of that which has transformed me. I, whose thought
> Is like a ghost shrouded and folded up
> In its own formless horror. Of all words,
> That minister to mortal intercourse,
> Which wouldst thou hear? For there is none to tell
> My misery . . .

>   (107–14)

What is compelling is the way these lines convey a sense of 'formless horror' by austerely declining to luxuriate in images. Beatrice's inability to confront her anguish is brought out; so, too, is her

awareness of this inability. And yet the writing is not so self-denying
that it is unable to suggest the crumpling of usual modes of thought,
the death of a stable sense of reality that leaves thought 'a ghost
shrouded and folded up | In its own formless horror'. The speech
couples self-assertion—'I, who', 'I, whose'—with a benumbed
awareness that the self being asserted has suffered violation.

Yet, if there is a continual effort to represent to the mind's eye
what is sensed as lying beyond representation, there is at least one
powerful stretch of writing where the otherness of the unconscious
is evoked as well as respected. This is Beatrice's 'description of the
chasm appointed for her father's murder' in III. i., a speech which
could be regarded, Shelley allows, as 'mere poetry' (*PP* 241):

>   there is a mighty rock,
> Which has, from unimaginable years,
> Sustained itself with terror and with toil
> Over a gulph, and with the agony
> With which it clings seems slowly coming down;
> Even as a wretched soul hour after hour,
> Clings to the mass of life; yet clinging, leans;
> And leaning, makes more dark the dread abyss
> In which it fears to fall: beneath this crag
> Huge as despair, as if in weariness,
> The melancholy mountain yawns . . . below
> You hear but see not an impetuous torrent
> Raging among the caverns, and a bridge
> Crosses the chasm; and high above there grow,
> With intersecting trunks, from crag to crag,
> Cedars, and yews, and pines; whose tangled hair
> Is matted in one solid roof of shade
> By the dark ivy's twine. At noon day here
> 'Tis twilight, and at sunset blackest night.
>
>   (247–65)

However, this admirable speech is, as Keach argues, 'anything but
an intrusion of . . . "mere poetry"'.[20] The subjective dimension—
the relevance of the passage to the spiritual states of both Cenci and
Beatrice—is strong enough to prevent us from reading the lines as
a piece of 'isolated description' (*PP* 241). Arguably, Shelley overdoes
the applicability of landscape to state of mind in the drawn-out

---

[20] Keach, 64.

comparison between the 'mighty rock' and 'a wretched soul', though Keach is right to speak of 'a curiously dizzying interpenetration of "imagery" and "passion"'.[21] But what enlivens the speech is Beatrice's movement from the 'undistinguishable mist | Of thoughts' (170–1) that tortured her earlier in the scene to the seemingly objective account of the chasm. The heroine's haunted cadences (especially evident in the last line and a half) momentarily close the gap between thought and language. Beatrice's account of the outer world's 'shade', 'twilight', and 'blackest night' suggests her sense of an inner 'dread abyss'. Apparent self-forgetfulness makes possible the emergence of a deeper awareness of self.

If here the gap between thought and language narrows, elsewhere the rift between Shelley's critique of his play and the play itself widens. The critic who writes the Preface preserves a distinction between essence and circumstance by reading the portrait of Beatrice in these terms: 'her nature was simple and profound. The crimes and miseries in which she was an actor and a sufferer are as the mask and the mantle in which circumstances clothed her for her impersonation on the scene of the world.' (*PP* 242.) Shelley's metaphors of 'impersonation' and 'scene' suggest a well-defined split between 'nature' and 'circumstances', essence and event. It is a split which, in Wasserman's view, is borne home by the dramatist's use of 'sublime casuistry', a casuistry that 'teaches us that evil is, though very real and persistently threatening, an "accident" of human nature, not its essence, and that it can be overcome by willful patience and endurance'.[22] But the ascent from 'restless and anatomizing casuistry' to 'sublime casuistry' is one which only those who choose to read the play in the light of some overarching system espoused by the poet will be able to make. Others will be more impressed by Shelley's convincing portrait of a person 'violently thwarted from her nature' (*PP* 238) who represses guilt and asserts an innocence that the onlooker can neither wholly validate nor deny.

In fact, in the play itself, Shelley casts a critical eye on the metaphors his characters use. Orsino's soliloquy in V. i. is a case in point:

> I thought to act a solemn comedy
> Upon the painted scene of this new world,
> And to attain my own peculiar ends
> By some such plot of mingled good and ill

[21] Keach, 65.   [22] Wasserman, 125.

As others weave; but there arose a Power
Which graspt and snapped the threads of my device
And turned it to a net of ruin . . .

(77–83)

Orsino still manœuvres within a polished, ironic rhetoric as he discards one illusion for another: the idea of pursuing a 'plot of mingled good and ill', of playing a part in 'a solemn comedy', gives way to the belief that his new disguises 'must be the masks of that within, | Which must remain unaltered' (V. i. 92–3). His subsequent resistance to an inadequate version of the self ('Shall I be the slave | Of . . . what? A word? (98–9)) suggests an inner vacuum more frightening than that sensed by the character himself. Paradoxically, what he calls 'that within' can be known only by way of the character's awareness of his lack of integrity.

In chapter seven of *The Marble Faun* Hawthorne uses the portrait of Beatrice Cenci as the point of departure for this exchange between Hilda and Miriam about Beatrice's sinfulness or otherwise:

'Sorrow so black as hers oppresses very nearly as sin would,' said Hilda.
'Then,' inquired Miriam, 'do you think that there was no sin in the deed for which she suffered?'
'Ah!' replied Hilda, shuddering. 'I really had quite forgotten Beatrice's history, and was thinking of her only as the picture seems to reveal her character. Yes, yes; it was terrible guilt, an inexpiable crime, and she feels it to be so. Therefore it is that the forlorn creature so longs to elude our eyes, and forever vanish away into nothingness! Her doom is just!'
'Oh, Hilda, your innocence is like a sharp steel sword!' exclaimed her friend. 'Your judgements are often terribly severe, though you seem all made up of gentleness and mercy. Beatrice's sin may not have been so great: perhaps it was no sin at all, but the best virtue possible in the circumstances. If she viewed it as a sin, it may have been because her nature was too feeble for the fate imposed upon her.'[23]

The two women's contrasting views mirror the contradictory responses engendered in the audience by *The Cenci*. The viewer or reader experiences both Hilda's orthodox recoil from sympathy into judgement, her sense that Beatrice's 'doom is just', and Miriam's rebellious surmise that Beatrice's 'sin' was 'no sin at all, but the

---

[23] *The Marble Faun* (New York, Scarborough, Ont., and London, 1961), 55.

best virtue possible in the circumstances'. Shelley arouses such
contradictory responses skilfully.[24]

It is towards the end of the play that what Joseph Donohue calls
'the blind partiality of Beatrice's view'[25] is experienced most keenly,
as in her response to the information that her mother and brother
have broken under torture:

> O white innocence,
> That thou shouldst wear the mask of guilt to hide
> Thine awful and serenest countenance
> From those who know thee not!
>
> (V. iii. 24-7)

The image of the 'mask' lends dualistic comfort to Beatrice who
clearly exculpates herself in these lines. The distinction between
'white innocence' and the 'mask of guilt' strikes one as self-deceivingly
casuistical. And yet here the spectator or reader is out on his or
her own, at once aware of Beatrice's sense of righteousness and
recognizing that, for her, there is no other course to take. It is not
surprising that Beatrice, caught between obedience to forms of
authority whose tyrannical nature has been made explicit and the
impulse to transgress, should feel so strongly that language falsifies,
something brought out by the echo of Iago's final words which
Shelley asks us to hear in this line: 'No other pains shall force another
word' (V. iii. 89). The fact that this is by no means Beatrice's final
speech sharpens our sense of her dilemma: mistrusting language,
unable to do without it.

Although awareness of self-division, of a 'suspected self' (II. ii.
102) such as troubles Giacomo, seems alien to Beatrice, the play
offers hints of its buried existence. In this speech she is subtly alive
to the deviousness with which Orsino's mind works:

> Even now you look on me
> As you were not my friend, and as if you

[24] Jeffrey Meyers contends persuasively that 'In *The Marble Faun* Hawthorne
exhibits the same defects as Guido Reni, who ignores Beatrice's crime . . . , and
impoverishes his art by effectively eliminating the "contact with sin" ', *Painting and
the Novel* (Manchester, 1975), 17. By contrast, I would argue, Shelley enriches his
art by highlighting his heroine's 'contact with sin'.
[25] 'Shelley's Beatrice and the Romantic Concept of Tragic Character', *KSJ* xvii
(1968), 71.

Discovered that I thought so, with false smiles
Making my true suspicion seem your wrong.

(I. ii. 30–3)

'Wrong' is pinned down as Orsino's self-pity by the scrupulous fixity
of 'false' and 'true'. Beatrice is characterized by her language here
as perceptive yet confined within an unyielding view of behaviour.
Her subsequent suggestion that she is in touch with a true self is not
without complacency: 'sorrow makes me seem | Sterner than else
my nature might have been' (I. ii. 34–5). And yet it is the self-
violence involved in that 'might have been' (and explored by the play)
which prompts sympathy for Beatrice.

That said, *The Cenci* questions as well as endorses Donohue's
reading of these lines: 'untoward circumstance is wrenching her out
of her true nature'.[26] Does the stress of tragic experience distort or
clarify a character's 'true nature'? Underscoring her belief in her
'subtle, pure, and inmost spirit of life' (III. i. 23), Beatrice assaults
abstractions which would codify it. The 'higher reason' (III. i. 363)
and 'higher truth' (V. ii. 164) which other characters see as justifying
the parricide both incite and frustrate the viewer's impulse to scent
out self-deception. We can sympathize with Beatrice's sense that
'truth', as defined by her Judges, is uncompromising and grotesque,
and we recoil from the First Judge's conviction that the rack can be
its mouthpiece:

> Dare you, with lips yet white from the rack's kiss
> Speak false? Is it so soft a questioner,
> That you would bandy lover's talk with it
> Till it wind out your life and soul?

(V. ii. 8–11)

Shelley's concern in this context with language and the self is
strikingly obsessive. That this parody of 'lover's talk' which the Judge
brutally imagines can have anything to do with 'life and soul' is a
horror to which Beatrice replies in riddles; she attempts to overthrow
the dominant language of authority. But it is doubtful whether her
equivocal, self-righteous language is much more authentic. When
Lucretia pleads with Beatrice—'O, speak the truth!' (V. iii. 55)—
her cry's despairing honesty cruelly exposes Beatrice's juggling with

---

[26] 'Shelley's Beatrice and the Romantic Concept of Tragic Character', 64.

words such as 'falsehood' (see V. iii. 51). Yet Shelley suspends his heroine in a linguistic limbo. By this stage in the play the word 'truth' has taken on complex and ironic colouring. Beatrice whose first words were 'Pervert not truth' (I. ii. 1) has ended up doing just that in order to resist the perversion of truth for which, in her view, her Judges are responsible.

Though no repose can be found inside Beatrice's consciousness, the play will not let the audience escape from it. It is fitting that one of Beatrice's final speeches (V. iv. 96–120) should address itself in a rhetorical, entrapped way to a word, 'hope'. Speech and consciousness give the impression of being restricted by language:

> Worse than despair,
> Worse than the bitterness of death, is hope:
> It is the only ill which can find place
> Upon the giddy, sharp and narrow hour
> Tottering beneath us.
>
> (V. iv. 97–101)

Beatrice is wilfully choosing the certainty of hopelessness, declining a Promethean belief in 'Love' which 'from the last giddy hour | Of dread endurance, from the slippery, steep, | And narrow verge of crag-like Agony, springs | And folds over the world its healing wings' (*Prometheus Unbound*, IV. 557, 558–61). To say that the speech is oratorical rather than inward is not to deny pathos to Beatrice's resolute refusal to entertain hope, but to suggest that the language prompts us to feel towards the speaker what Stuart Sperry calls a 'delicate balance of sympathy and ironic awareness'.[27]

Elsewhere, Shelley wrongfoots his audience by making Beatrice speak with a belief in her guiltlessness which seems close to delusion (as in, for example, her concluding speech in IV. iv. where she speaks of 'triumphant Innocence' (184) as though in relation to herself). The 'restless and anatomizing casuistry' that Shelley discusses in his Preface is again prompted, and indeed the play's treatment of moral and emotional contradictions threatens to undercut his claim to teach 'the human heart, through its sympathies and antipathies, the knowledge of itself' (*PP* 240). The existence of so sturdy and unchanging an absolute as 'the human heart' is questioned by a play that provokes us to rethink our 'sympathies and antipathies'.

[27] 'The Ethical Politics of Shelley's *The Cenci*', *SIR* xxv (1986), 427.

Beatrice's famous speech earlier in V. iv. 47–75 suggests one kind of imaginative authority which *The Cenci* possesses. One may deny finished control to the lines but not the sudden thrust of nightmare:

> If there should be
> No God, no Heaven, no Earth in the void world;
> The wide, grey, lampless, deep, unpeopled world!
> If all things then should be . . . my father's spirit . . .
>
> (V. iv. 57–60)

Several echoes, 'unpeopled' and 'lampless' in particular, allow one to track down the imaginative stimulus here as being similar to that which informs Panthea's vision: ' . . . and every space between | Peopled with unimaginable shapes | Such as ghosts dream dwell in the lampless deep' (*Prometheus Unbound*, IV. 243–5). Yet *The Cenci* parodies the later visionary wonder. Shelley moves from self-reflexive intricacies into the power of Beatrice's obsession that God might be supplanted by Cenci. However, the writing's pitch is both difficult and undesirable to sustain. Only so long as the obsession seems the product of unconscious forces does it compel. Certainly, Beatrice's lines conclude awkwardly with an ill-concealed borrowing from Hamlet's best-known soliloquy: 'Who ever yet returned | To teach the laws of death's untrodden realm?' (72–3). But her speech undermines religious belief in an ultimate sanctuary which, paralleling the belief in an inviolable inner self, has glimmered as a possible consolation throughout the play.

The much-admired conclusion offers the balm of a not uncritical pathos:

> Give yourself no unnecessary pain,
> My dear Lord Cardinal. Here, Mother, tie
> My girdle for me, and bind up this hair
> In any simple knot; aye, that does well.
> And yours I see is coming down. How often
> Have we done this for one another; now
> We shall not do it any more. My Lord,
> We are quite ready. Well, 'tis very well.
>
> (V. iv. 158–65)

This is inward rather than oratorical, even if the conversational tone is aware of itself as the product of certain stylizing effects: simplicity of diction, the varying position of caesurae. In other words, the writing has its eye too much on itself to be straightforwardly moving, yet the reader's hesitation is calculated by Shelley. We are invited to admire guardedly the stoical courage displayed in Beatrice's final words. The lines point a marked contrast between the self-steeled but arguably self-deceived Beatrice of the end and the Beatrice who experiences a sense of alienation from self in the immediate aftermath of the rape:

> How comes this hair undone?
> Its wandering strings must be what blind me so,
> And yet I tied it fast.—O, horrible!
> The pavement sinks under my feet! The walls
> Spin round! I see a woman weeping there,
> And standing calm and motionless, whilst I
> Slide giddily as the world reels . . .
>
> (III. i. 6–12)

The passage from 'undone' hair to bound-up hair mirrors the heroine's change from trauma to stoicism. Yet the earlier lines come closer to the character's centre of self (or fear that such a centre has been violated). For what makes The Cenci arresting is its questioning of beliefs (both orthodox and Shelleyan), the ruthless drive of its pessimism, its fascination with the 'formless horror' (III. i. 111) that coils inside 'deserted selves' (V. iii. 69). It is undoubtedly the case that The Cenci incites contrasts with Prometheus Unbound. But the temptation to read the play in the light of the lyrical drama should be resisted whenever it leads the reader away from the impact and power of the individual work. Stuart Curran's remark that 'the universal corruption of The Cenci appears flatly to contradict the universal benevolence of Prometheus Unbound'[28] is where readings of the play should end as well as begin. Both works explore 'what a world we make, | The oppressor and the oppressed' (The Cenci, V. iii. 74–5), but it is Beatrice's response to oppression which explains the utterly different vision of The Cenci. Characters and audience are tangled in a casuistical, equivocating net. Yet if the play's pessimism is hard to ignore, so, too, is its power to stir and involve. It is to Prometheus Unbound that a reader in search of a more hopeful, idealizing poetry must turn.

---

[28] Shelley's Annus Mirabilis, 121.

5

Prometheus Unbound
A Perpetual Orphic Song

ACT I

The ideological drift of *Prometheus Unbound* may seem evident. However, throughout the work the reader is aware of a drive to complicate and enrich meaning. Angus Fletcher sees such a drive as characteristic of works which occupy 'that borderland where one passes from allegory into myth';[1] certainly, the way the lyrical drama operates makes 'allegory' a term of only fitful relevance. The inventiveness of Shelley's fiction-making results in a subtle and self-aware poem, one that is most compelling when least inclined to propagandize. Indeed, the most serious risk the poem runs is of laying claim to an imaginative autonomy which is self-enclosed: there are occasions when its lines ask too undemandingly to be read 'not as pointers towards an external reality but as constituting reality itself'.[2] But the poem often ceases to be a straightforward vehicle for ideas when it is operating most imaginatively.

Marlon Ross both tries to resolve the major critical problem posed by *Prometheus Unbound* and hints at the source of its poetic achievement when he contends that in the poem 'Shelley's objective . . . is to lodge within the mind relations so complex that they cannot be reduced to already apprehended relations'.[3] Only thus, he suggests, can Shelley shake the reader out of habitual responses and prepare the way for the mental liberation which the poet desires. His article is alive to the far-reaching implications of Shelley's attempt to find an idiom which would avoid the didactic (the poet's 'abhorrence', *PP* 135) but 'familiarise the highly refined imagination

---

[1] *Allegory: The Theory of a Symbolic Mode* (Ithaca, NY, 1964), 322.
[2] Geoffrey Ward, 'Transforming Presence: Poetic Idealism in *Prometheus Unbound* and *Epipsychidion*', *EOS* 198.
[3] 'Shelley's Wayward Dream-Poem: The Apprehending Reader in *Prometheus Unbound*', *KSJ* xxxvi (1987), 115.

of the more select classes of poetical readers with beautiful idealisms of moral excellence' (*PP* 135). In place of ladling out propagandist directives, the poem, according to Ross, prompts the reader to experience a recurring 'psychological effect of gain and loss, fulfillment and intensified desire, expansion and emptiness'.[4] Yet, persuasive as this is in many ways, it too blandly harmonizes the tensions, pressures, and contradictions which give what is best in the work vitality and interest. Arguably, it is when the poem proceeds with an over-sure sense of where it wishes to go that it risks becoming glib or predictable.

As Susan Brisman points out: 'Against his habitual concern about the inconstancy of the wind of inspiration, the poet of *Prometheus Unbound* seems imaginatively capable of endlessly inventing fictions that defer redemption.'[5] Her wording pinpoints a rewarding tussle between the lyrical drama's overall purpose ('redemption') and its stage-by-stage inventions ('fictions'). I take the tricky word 'fictions' to imply uses of language that are less interested in dutifully earning their thematic keep than in pursuing, in a more localized, self-sufficient way, possibilities thrown up by the imagination. An example from Act I is Earth's imagining of a shadow world:

> They shall be told.—Ere Babylon was dust,
> The Magus Zoroaster, my dead child,
> Met his own image walking in the garden.
> That apparition, sole of men, he saw.
> For know, there are two worlds of life and death:
> One that which thou beholdest, but the other
> Is underneath the grave, where do inhabit
> The shadows of all forms that think and live
> Till death unite them, and they part no more;
> Dreams and the light imaginings of men
> And all that faith creates, or love desires,
> Terrible, strange, sublime and beauteous shapes.
>
> (191–202)

These fine lines pose any reader of the poem with a crucial test-case. Responding to Prometheus' plea, 'But mine own words, I pray, deny me not' (190), Earth has access to a vision that flowers out of the

---

[4] 'Shelley's Wayward Dream-Poem', 126.
[5] ' "Unsaying His High Language": The Problem of Voice in *Prometheus Unbound*', *SIR* xvi (1977), 86.

hero's sense of being separated from his previous speech. Yet the
shadow world outgrows the function it is invented to perform (that
of housing a phantom who can be called forth to 'utter | The curse
which all remember' (209–10)). There is a significant gap between
'They shall be told' and the words that follow; it is Earth's surrender
to her hitherto 'unattempted'[6] vision that gives the passage authority
and distinguishes it from Byron's derivative passage in *Cain*.[7] Those
'shadows of all forms that think and live' tease us out of thought.
The reader moves towards but never reaches a rationally explicable
position. Are there two parallel worlds of life and death? Is the world
which 'thou [Prometheus] beholdest' our world? How hard can the
Platonic suggestions of 'shadows' and 'forms' be pressed? Though
such questions arise, it is our inability to answer them that illuminates
Shelley's achievement. For all the accents of exposition—'For know,
there are two worlds of life and death'—what Shelley offers is
phantasmagoria. The following lines contribute to this effect: 'Dreams
and the light imaginings of men | And all that faith creates or love
desires'. If the first line deprecates 'light imaginings' of which Shelley's
entire poem is an example, the second line repudiates the deprecation.
Imagination reaches its vanishing point in the culminating line,
'Terrible, strange, sublime and beauteous shapes'; the shadow world
might be a repository of the contradictory possibilities of human
experience, its coexisting but contrasting 'shapes' both created by
and a riddle to the poet. There is, in the following lines that people
the shadow world with the dramatis personae, a distinct sense of
moving into a realm which is hardly distinguishable from the inner
world of the imagination, of being taken behind the scenes of the
drama. There is, too, a slight sense of strain as the poet attempts
to weave his isolated imagining into his overall design:

> There thou art, and dost hang, a writhing shade
> 'Mid whirlwind-peopled mountains; all the Gods
> Are there, and all the Powers of nameless worlds,

[6] See Shelley's account of his work as 'a drama, with characters & mechanism of
a kind yet unattempted', *Letters*, ii. 94.

[7] See *Cain*, II. i. 151–6, where the 'phantasm of the world' (152) is described by
Leslie Brisman as 'a Luciferic reduction to the level of clap-trap of Shelley's magnificent
fiction of a shadow world in *Prometheus Unbound*', though he goes on to contend
that the 'belatedness of this world has an authentic ring', 'Byron: Troubled Stream
from a Pure Source', *ELH* xlii (1975), 625.

Vast, sceptred phantoms; heroes, men, and beasts;
And Demogorgon, a tremendous Gloom . . .

(203-7)

The whole passage could be viewed as offering an image of
the mental division which is a central concern of the act.[8] But
whether the passage can or should be thematically pigeon-holed is
questionable. Shelley's vision replaces habitual 'relations' with an
imaginative space, a 'making strange' that one might see as preparing
the way for awakened thoughts of liberation. Yet such a reading
would be merely one response to the rich enigma of the text.
Prometheus Unbound's fineness as a poem has less to do with the
coherence of the myth it tries to construct than with its readiness
to admit that it is involved in making myth, its ability, in Brian
Nellist's words, 'to accommodate contradictions and acknowledge
uncertainties'.[9] That the shadow world contains 'Dreams and the
light imaginings of men' both gestures towards and ironizes the
drama's own origins. And that the shadow world contains the
dramatis personae momentarily brings to light the poem's sense of
itself as a projection of thought. There are for Shelley political and
moral implications bound up in the recognition that reality is what
we make of it: we can choose to see as Jupiter sees, or as Asia and
Prometheus see. But it is more the imaginative recognition than the
moralized implications which the work brings to our attention. What
Shelley dramatizes throughout is the unavoidability of interpretation.
Characters are shown choosing to see or read reality in a certain light
and, as a consequence, shaping reality. The same burden and the
same freedom are proffered to the reader by a text that is endlessly
generous in possibilities of interpretation. It is, then, the ability of
the shadow world to resist our desire for unequivocal meaning which
makes it an apt emblem of the imaginative openness of mind that
Shelley wishes to encourage.

However, the poem's attempt to suggest new 'relations' of thought
by employing a system of cross-referencing echoes can suggest covert
anxiety about the work's overall coherence. As is the case in parts
of *Epipsychidion* (see Chapter Seven), Shelley's wish to bind together

---

[8] This is the view taken by Richard Cronin who argues that 'The mind is divided.
It has rejected a part of its own substance, accrediting to that part an independent
reality', Cronin, 137.
[9] 'Shelley's Narratives and "The Witch of Atlas"', *EOS* 173.

can lead to a loss of particularity. Indeed, the way that the poem constantly turns in upon itself is the source both of strength and weakness, subtlety and over-subtlety. As William Hildebrand puts it: 'Press *Prometheus Unbound* almost anywhere, one is sometimes moved to think, and one can sense the shape of that intricate trellis of mutually inflecting metaphors and symbols that constitutes Shelley's ontological and epistemological intuitions in the drama.'[10] That the patterning of such intuitions has its own beauty is undeniable. But the poetry can seem self-absorbed, entangled in its own finely spun webs as in these lines from Act I when Earth summons the Spirits of the Human Mind to come to the hero's aid:

> To cheer thy state
> I bid ascend those subtle and fair spirits
> Whose homes are the dim caves of human thought
> And who inhabit, as birds wing the wind,
> Its world-surrounding ether; they behold
> Beyond that twilight realm, as in a glass,
> The future—may they speak comfort to thee!

(657–63)

Giving a local habitation and a name to the conceptual, Shelley is at once precise and elusive. The passage repays analysis without escaping a slightness intrinsic in its very fluency. What appears to open into illumination ends up involved within itself. For instance, 'world-surrounding ether' cleverly but patly blurs the careful segregation of inner and outer on which the simile, 'as birds wing the wind', seems intent. Shelley's images erase distinctions in order to suggest the all-pervasiveness of human thought. But it is one thing to be able to offer a paraphrase of a passage, another to feel that meanings are fully realized in the language. It is possible to respect the quickness of mind displayed by the poetry, yet to feel that the passage leaves the reader with a disembodied nuance, the ghost of a meaning. Sometimes, as here, Shelley expresses a crucial theme with a riskily airy lightness. A further example is the Fourth Spirit's song, which, for 'the *locus classicus* . . . of the passionate defence of visionary poetry in English',[11] might seem to articulate its theme with disabling ease. The more cavalier imaginings of *The Witch of Atlas*, as I shall

---

[10] 'A Look at the Third and Fourth Spirit Songs: *Prometheus Unbound, I* ', *KSJ* xx (1971), 87.
[11] *SM* 115.

argue in Chapter Six, treat such 'Nurslings of immortality' (749) with a comic, mocking bravura which, paradoxically, makes them more available to the reader's imagination.

Prometheus Unbound, like *The Cenci*, is often most intriguing when it places in the foreground the struggle to articulate its vision. In the lyrical drama the writer's struggle to create is linked with his characters' struggle to win freedom from habitual modes of perception. So, in the exchange between Earth and Prometheus in Act I, the play's invented reality is made strange by Earth's insistence that Prometheus cannot hear:

> No, thou canst not hear:
> Thou art immortal, and this tongue is known
> Only to those who die . . .
>
> (149–51)

However, Shelley's treatment of language and its deceptions can result in writing which is as shadowy as it is subtle, as in the Phantasm of Jupiter's self-consciousness:

> What unaccustomed sounds
> Are hovering on my lips, unlike the voice
> With which our pallid race holds ghastly talk
> In darkness?
>
> (242–5)

The irony implicit in the fact that the Phantasm is called up to speak the curse pronounced by the pre-play Prometheus is apparent. The act is strewn with hints that Jupiter is the hero's double, a shadow self created (or miscreated) by his imaginings. But here we must wait for Prometheus' response—'Speak the words which I would hear, | Although no thought inform thine empty voice' (248–9)—before enigma is given some semblance of purpose, as Susan Brisman brings out: 'Promethean voice is always reclaiming speech from a language of reference, where words are assumed to have stronger relations to the objects and thoughts they represent than to one another.'[12] Prometheus' request for speech unmoored from consciousness plays an ironic variation on this theme. More generally, Brisman's statement suggests a rationale for the poetry's attempt to subvert

---

[12] ' "Unsaying His High Language" ', 59.

allegorical content: an escape from allegory emerges as a prerequisite
of the freedom which is the drama's central concern.

For all its pursuit of such freedom, there are times in the first act
when *Prometheus Unbound* seems to believe only partially in its own
created world. Such partial belief is not unconnected with, but results
in less satisfactory poetry than, the lyrical drama's impulse to
complicate its impact. Although the poem's reworkings of Aeschylus
and Milton are intelligent and purposeful, they betray indebtedness
to the earlier writers' wording and inventive powers. In Prometheus'
opening speech, for example, Shelley's revisionary intelligence
threatens to overwhelm his poem's imaginative identity; it is as
though the poet had one eye on an open text of *Prometheus Bound*
in lines such as these, adapted from Aeschylus,[13] but given to
Prometheus to emphasize his 'triumphant stoic defiance of pain
through faith in the inevitable workings of Necessity':[14]

> And yet to me welcome is Day and Night,
> Whether one breaks the hoar frost of the morn,
> Or, starry, dim, and slow, the other climbs
> The leaden-coloured East; for then they lead
> Their wingless, crawling Hours . . .
>
> (44–8)

The writing shoulders its burden of significance too painstakingly
for its own creative good. And Shelley's attempts to represent recent
and contemporary history (the French Revolution and its aftermath)
substitute melodramatic assertion—'See how kindred murder kin!
| 'Tis the vintage-time for Death and Sin' (573–4)—for the kind
of vivid imaginative presence achieved by Georg Büchner, say, in
Danton's Death.[15] Through the Furies Shelley implies an interesting
view of the Revolution: that its initial impulse towards 'Freedom'

---

[13] The lines from *Prometheus Bound* which Shelley adapts here are lines 23–7,
spoken by Hephaestus and translated by Philip Vellacott in this way: 'Glad you will
be to see the night | Cloaking the day with her dark spangled robe; and glad | Again
when the sun's warmth scatters the frost at dawn. | Each changing hour will bring
successive pain to rack | Your body', '*Prometheus Bound*', '*The Suppliants*', '*Seven
Against Thebes*', '*The Persians*', trans. Philip Vellacott (1961; Harmondsworth, 1970),
21.

[14] Wasserman, 286.

[15] See my 'A More Hazardous Exercise: Shelley's Revolutionary Imaginings',
forthcoming in J. R. Watson (ed.), 'The French Revolution in English Literature and
Art: Special Number', *The Yearbook of English Studies*, xix (1989), for a fuller
discussion of Shelley's treatment of the French Revolution.

(570) was wrecked by the Terror, that the consequent 'Despair' (576) made inevitable a return to the vicious circle of 'slaves and tyrants' (577). But he does so in language which is less than fully interesting, a language of hectic statement and political shorthand.

Yet the insistence on the supreme importance of interpretation is, as I have argued, a vital strength of the lyrical drama; and though the poetry cannot match Wordsworth's emphasis on lived experience in his account of the Revolution in *The Prelude*, it comes to life when it focuses on the effect on Prometheus of the Furies' temptations to despair. Moreover, the language draws attention to the fact that the theatre in which the Revolution is being re-enacted is mental: 'Tear the veil!' (539), commands the Fury, and what follows stages the inner conflicts of a generation of disillusioned liberals. But the writing in this passage (567–615) lacks the creative energy of the best parts of *Prometheus Unbound* (the Fury's great speech at line 618, discussed below, is a different matter). The lines sacrifice the visionary power they seem intent on in order to drive home their polemical point; their determination to offer, in the Fury's word, an 'emblem' (594) of what Shelley wants us to see as mistaken disenchantment is too transparent.

Certainly, the first act is compelling when its language links 'the operations of the human mind' with 'those external actions by which they are expressed' in such a way that distinctions between the 'internal' and 'external' still obtain.[16] For instance, the reader can see the Furies as embodiments of the dependent but separate life taken on by conceptual products of the mind:

SECOND FURY

The beauty of delight makes lovers glad,
Gazing on one another—so are we.
As from the rose which the pale priestess kneels
To gather for her festal crown of flowers
The aerial crimson falls, flushing her cheek—
So from our victim's destined agony
The shade which is our form invests us round,
Else are we shapeless as our Mother Night.

(465–72)

[16] See Shelley's account in the Preface of the sources of his 'imagery': 'The imagery which I have employed will be found in many instances to have been drawn from the operations of the human mind, or from those external actions by which they are expressed', *PP* 133.

The use of simile is an admission of the ungraspable; Shelley exacerbates the clash between the incorporeal and the image. This description of the way the Furies exist parodies the account of language as Orphic song in Act IV (415–17), and makes them disquieting reflections of the 'shapes' which haunt the drama's 'wildernesses' (742) in Act I. Wasserman's account of Shelley's scepticism about language's relationship with reality is relevant: 'Shelley's rejection of materialism and of the dualism of subject and object must have driven him to reconsider the function of language, for he could no longer assume it to be an analysis of percepts into the components and relationships obtaining among their counterparts in an outside reality.'[17] But what makes the writing successful as poetry is the way it both allows for and subtly negates 'the dualism of subject and object'. So Prometheus' disgust, 'execrable shapes' (449), is set against his inability to identify his tormentors, 'Horrible forms, | What and who are ye?' (445–6). This inability prepares the reader for the taunting way the Third Fury defines what the Furies are in lines 483–91, their identity withheld even as the Fury puts into words what Prometheus thinks they are. By refusing final commitment to its own words, the Fury acts as a demonic counterpart of the visionary poet of Act I. Just as there is a tug between image and abstraction in the presentation of the Furies, so throughout Act I the presence of ideas and feelings that resist definition both interferes with the poetry's imaginative life and contributes to its finest effects.

The best poetry in the first act brings out the difficulty and inwardness of Prometheus' struggle, and gives the lie to any account of the play that insists on the *naïveté* of its vision. The lyrical drama may be ultimately hopeful, but only after it has taken a long look at the worst. The Fury's speech to Prometheus beginning at line 618, for example, obliges one to question the correctness of Yeats's assertion that Shelley 'lacked the Vision of Evil':[18]

> The good want power, but to weep barren tears.
> The powerful goodness want: worse need for them.
> The wise want love, and those who love want wisdom;
> And all best things are thus confused to ill.

[17] Wasserman, 267.
[18] In *A Vision* (1937; London, 1981): 'He [Shelley] lacked the Vision of Evil, could not conceive of the world as a continual conflict, so, though great poet he certainly was, he was not of the greatest kind', p. 144.

Many are strong and rich,—and would be just,—
But live among their suffering fellow men
As if none felt: they know not what they do.

(625–31)

Though the play attempts to transcend this vision of contradictions, it is no accident that the speech is among the most achieved passages in *Prometheus Unbound*, its barely tolerable paradoxes spotlighted by rhythms and diction that are clear yet troubled. The language suffers its insights; Shelley's analysis of impasse at once sobers and focuses his imagination. Indeed, the poetry's sensitivity to the predicament of the individual torn by unresolved doubts and complicated feelings pervades Act I. It can be seen at work in Prometheus' opening speech with its authentically unconvincing repression of antagonism towards Jupiter—'Disdain? Ah no! I pity thee' (53)—as well as its suggestion of the hero's almost narcissistic fascination with his own suffering. Lines 31–3 ('The crawling glaciers pierce me with the spears | Of their moon-freezing chrystals; the bright chains | Eat with their burning cold into my bones') absorb pain into a labyrinth of assonance, and are used by Peter Conrad to support the generalization that 'The romantics subvert drama by making it lyrical'.[19] Here, however, the lyrical subversion of drama is itself dramatic. Equally able to insinuate criticism of narcissism is the Sixth Spirit's haunting lyric, 'Ah, sister! Desolation is a delicate thing' (772); the song uses lulling, undulating rhythms with hard-headed awareness of the insidious attractions of 'Desolation', the condition of disillusion that tempts the 'best and gentlest' to forsake their 'tender hopes' (775).

'How vain is talk!' (431), says Prometheus to Panthea before the appearance of the Furies, pinpointing the mistrust of language which eloquently registers self-division in the poetry of Act I. In one sense 'talk' is 'vain' because the choices are clear-cut, dependent on the right orientation of the will; after his declaration of repentance at line 303 ('It doth repent me: words are quick and vain'), there is nothing for the hero to do but hold out. But, in another sense, 'talk' is the treacherous medium through which consciousness must articulate its aspirations. It is the tension between these aspirations and the resistances to them which is embodied in the first act's

---

[19] *The Victorian Treasure-House* (London, 1973), 18.

language. The rhythm of Prometheus' penultimate speech is weighed
down by the sense of impasse which vitalizes the poetry:

> I would fain
> Be what it is my destiny to be,
> The saviour and the strength of suffering man,
> Or sink into the original gulph of things. . . .
> There is no agony and no solace left;
> Earth can console, Heaven can torment no more.
>
> (815–20)

The polemical bite of that final line, with its reversal of orthodox
valuations of Earth and Heaven, suggests one source of energy in
Prometheus' language. But a deeper source is the mirroring of the
hero's struggle by the poet's own struggle to create an appropriate
myth. The intermittent success of the poetry of Act I has to do with
the laying bare of this second struggle.

ACT II

Tortured by the Furies' 'emblem' (I. 594) and 'sights' (I. 643)
Prometheus prophesies an hour 'When they shall be no types of things
which are' (I. 645). The recovering world of Act II is described by
Shelley in a language that is strikingly autonomous, removed from
any emblematic idiom of 'types'. Much of the writing in Act II is
'haunted by sweet airs and sounds' (I. 830). Teeming with images
of uncertain existence, brief awakenings, hints and fadings and
collapses, the verse is absorbed by its own presence. 'How the notes
sink upon the ebbing wind!' (II. i. 195), says Asia, for instance. Such
writing may seem in danger of preciosity, yet the first speech of the
act—Asia's exultation that 'Spring' has come and her description
of the waning morning star—introduces the reader to the poetic
challenge offered by Shelley's concentration on the process of change:

> From all the blasts of Heaven thou hast descended—
> Yes, like a spirit, like a thought which makes
> Unwonted tears throng to the horny eyes
> And beatings haunt the desolated heart
> Which should have learnt repose,—thou hast descended
> Cradled in tempests; thou dost wake, O Spring!
> O child of many winds!
>
> (1–7)

'Thou dost wake'; the rest of the act prolongs the process of awakening. In the passage above, the final exclamation develops the suggestion (floated by the previous line's paradoxical 'Cradled in tempests') of peace won from, and in the midst of, struggle. Moreover, simile surprises the poetry into greater depth. Although it may belong to Asia, 'the desolated heart | Which should have learnt repose' has a more general impact. At times in Act II, as in the account of the waning star (II. i. 17–27), the writing prompts a conflict of responses: admiration for its delicacy, grace, and lyricism; concern that it is growing self-engrossed, decorative. But the achievement is real though it falls short of Shelley's greatest poetry.

This act, too, is aware of the problems of 'voice', of finding an appropriate idiom. Occasionally such awareness displays itself in 'a surfeit of energy vexing its own articulation'.[20] But more often such energy liberates rather than vexes, self-consciousness here being coupled with imaginative freedom. For the rest of the first scene of Act II, there is great assurance in the sequence of incidents. Lyric detail finely supports the overall development, as in the description of the clouds being 'Shepherded by the slow, unwilling wind' (II. i. 147). Arnold selected the line as illustrating the false beauties to which Shelley and Keats devoted themselves. And certainly there are times when Shelley does seem intent merely on the production of 'exquisite bits and images'.[21] But more than decoration is evident here. The line is, in fact, characteristic of this scene, not only in its rhythmic cunning—it buckles at 'slow' before regaining straggling momentum in 'unwilling'—but also in its connection with Asia's previous sense of wonder:

> As you speak, your words
> Fill, pause by pause my own forgotten sleep
> With shapes.

(141–3)

The 'continuous expression of slow movement'[22] which Ruskin praises the poetry for capturing reveals a wariness of emphasis, an attempt to convey the gradual emergence in nature and consciousness

---

[20] Susan Brisman's account of II. i. 50–1, ' "Unsaying His High Language" ', 79.

[21] See G. M. Matthews (ed.), Keats: The Critical Heritage (London, 1971), 326.

[22] Modern Painters, I (1903), The Works of John Ruskin, ed. E. T. Cook and Alexander Wedderburn (39 vols., London, 1903–10), iii. 364.

of states of hope and awakening. The poetry searches for states of suspension that hint at but refuse to spell out significance, as in lines from the same speech—

> And the white dew on the new-bladed grass,
> Just piercing the dark earth, hung silently
>
> (148–9)

—where the precision with which Shelley places 'Just' and 'hung silently' is suggestive.

Indeed, though the writing looks forward, full of anticipation, it is still able to relish, even spin out the process of change. Throughout this first scene Shelley achieves a steadied intensity. In Panthea's dream of union with Prometheus, for instance, the poetry attends to each moment of the experience and attains 'that tranquillity which is the attribute and accompaniment of power':[23]

> I saw not—heard not—moved not—only felt
> His presence flow and mingle through my blood
> Till it became his life and his grew mine
> And I was thus absorbed—until it past
> And like the vapours when the sun sinks down,
> Gathering again in drops upon the pines
> And tremulous as they, in the deep night
> My being was condensed . . .
>
> (79–86)

The work that Shelley persuades his rhythms to perform, especially the effect of prolongation at each line-ending, anticipates the greatest poetry in *Epipsychidion*. Words such as 'presence', 'became', 'grew', even the over-used 'felt', develop new possibilities as they are spoken. The scientific analogy with the processes of condensation tightens the writing, and the simile deepens our grasp of the experience.

That this account of a changed stage moves quickly over the crisis is evidence that Demogorgon's assertion—'the deep truth is imageless' (II. iv. 116)—is relevant to Shelley's language. Throughout Act II, the visual is overpowered by the visionary (the fine lyrics which conclude the fifth scene supply impressive examples); the poetry explores transformation, a process which often involves

---

[23] See letter to Godwin quoted by Mary Shelley: 'Yet, after all, I cannot but be conscious, in much of what I write, of an absence of that tranquillity which is the attribute and accompaniment of power', *P W* 158.

redefinition of the word 'see', as when Panthea attempts to describe Demogorgon:

> I see a mighty Darkness
> Filling the seat of power; and rays of gloom
> Dart round, as light from the meridian Sun,
> Ungazed upon and shapeless—neither limb
> Nor form—nor outline; yet we feel it is
> A living Spirit.
>
> (II. iv. 2–7)

'Notoriously hard to define':[24] Timothy Webb's description of Demogorgon is borne out and turned into a virtue by these lines. A calm assertion of Demogorgon's 'felt' reality is coupled with an equally calm denial that it can be seen. The simile of 'light from the meridian Sun' illuminates the paradox of a 'mighty Darkness' that Panthea sees even though it is 'Ungazed upon and shapeless'. Shelley gives poetic shape to his sceptical thinking about history and finds an idiom 'shapeless' enough to respect such scepticism. The slow-moving cancellations in the language offer proof that this 'living Spirit' enjoys authentic existence.

What, then, is the relationship between lyrical invention and visionary politics in Act II? In a fascinating attempt to marry the two, G. M. Matthews spells out the thematic implications of Shelley's images, especially images related to volcanic activity whose basis is said to be 'the perception of revolutionary activity in the external world and in the human mind—of irrepressible collective energy contained by repressive power'.[25] Such 'over-determined' concepts serve as 'a collecting-point for several of the writer's political, scientific, or philosophical perceptions of reality'.[26] There is a revealing gap between the cautious vagueness of this definition and the particularized richness of Matthews's readings; at a theoretical level Matthews wishes to avoid reductive allegorizing; at a practical level he illuminates contexts out of which the poetry has arisen. In his discussion of II. ii., his readings are persuasively relevant to a scene that talks of 'Demogorgon's mighty law' (43), and ends with the Second Faun's anticipation of a profounder poetry: 'those wise

---

[24] *Percy Bysshe Shelley: Selected Poems*, ed. Timothy Webb (London and Totowa, NJ, 1977), 197.
[25] 'A Volcano's Voice in Shelley', *ELH* xxiv (1957), 222.
[26] 'A Volcano's Voice in Shelley', 192.

and lovely songs | Of fate and chance and God, and Chaos old'
(91–2). However, Matthews solves the problem of texture by
concentrating on it with brilliant but limiting exactness, as is
illustrated by his account of the scene's first lyric, which is said to
draw on the associations of the over-determined concept (in this case
the volcano), and read as heralding revolutionary change: 'the lush
exuberance of the flora and fauna, the interwoven bowers and
voluptuous nightingales, have been criticized as excessive, but the
lushness is neither fanciful nor gratuitous. Asia and her companion
have reached an area of volcanic fall-out, long famous for extreme
fertility.'[27]

Developed to avoid the rigidity of the symbol, the over-determined
concept imposes its own kind of rigidity. The reader does not have
to believe in what Matthews satirizes as 'an incorruptible mystical
system'[28] to see that Shelley is writing about politics in the most
lyrically oblique way possible:

SEMICHORUS I OF SPIRITS

> The path through which that lovely twain
>   Have past, by cedar, pine and yew,
>   And each dark tree that ever grew
>   Is curtained out from Heaven's wide blue;
> Nor sun nor moon nor wind nor rain
>   Can pierce its interwoven bowers . . .

(1–6)

The third line needs to be read twice. As Wasserman points out,[29]
Shelley's pastoral refuge is united in the phrase, 'And each dark tree
that ever grew', to the phenomenal world. The stressing of 'each'
and 'ever' is uncompromising. The material world is present as an
absence of sun, moon, wind, and rain. Ensuing lines sustain this
evocation of two separate realms of existence, which the nymphs
will unite by their journeying; of glimmers of insight which will lead
ultimately to revelation:

> Or when some star of many a one
> That climbs and wanders through steep night,
> Has found the cleft through which alone
> Beams fall from high those depths upon,

---

[27] 'A Volcano's Voice in Shelley', 205–6.
[28] 'A Volcano's Voice in Shelley', 193.   [29] Wasserman, 314.

Ere it is borne away, away,
By the swift Heavens that cannot stay—
It scatters drops of golden light
Like lines of rain that ne'er unite;
And the gloom divine is all around
And underneath is the mossy ground.

(14–23)

The lyric does not so much act as the vehicle of a particular idea
as make the reader receptive to the fleeting entrance into history and
consciousness of such hopeful intimations as are suggested by the
'star' and its 'drops of golden light'. Here it is the poetry's calmly
non-logical intuition of a simultaneous absence and presence, of what
D. J. Hughes calls 'the gradual emptying-out of the phenomenal and
the suggested and gradual presence of the noumenal',[30] which is
impressive.

Angus Fletcher argues: 'Shelley wanted his readers to struggle
with the grand and confused scenes of his odes, with the obscure
metaphysics of his *Prometheus*, for which purpose he elaborated a
highly ornamental style.'[31] But Shelley's 'highly ornamental style'
serves both as a camouflage for 'metaphysics' and as a source of self-
justifying pleasure. The cunning, for instance, with which the third
semi-choral lyric in II. ii. presents the issue of free will and necessity
springs from a mind ready to stand at a remove from the dogmatic
conviction that change is, in some impersonal way, inevitable:

those who saw
Say from the breathing Earth behind
There steams a plume-uplifting wind
Which drives them on their path, while they
Believe their own swift wings and feet
The sweet desires within obey . . .

(51–6)

The Spirits describe the operation of 'Demogorgon's mighty law' on
those 'destined' (50) to be stirred from sleep (presumably into
awareness of the need for change). Yet the poetry allows for an
understanding of this mighty law that does not discount (even as
it does not wholly endorse) the involvement of the individual will.

---

[30] 'Potentiality in *Prometheus Unbound*' [*SIR* ii (1963), 107–26], in R. B.
Woodings (ed.), *Shelley: Modern Judgements* (London, 1968), 151.
[31] *Allegory: The Theory of a Symbolic Mode*, 277.

But the way the poetry's 'ornamental style' delights in its own fictiveness is evident in a speech towards the close of the scene; the writing suggests Shelley's discovery of a seemingly fanciful yet richly achieved way of discussing his own poetry:

> I have heard those more skilled in spirits say,
> The bubbles which the enchantment of the sun
> Sucks from the pale faint water-flowers that pave
> The oozy bottom of clear lakes and pools
> Are the pavilions where such dwell and float
> Under the green and golden atmosphere
> Which noontide kindles through the woven leaves,
> And when these burst, and the thin fiery air,
> The which they breathed within those lucent domes,
> Ascends to flow like meteors through the night,
> They ride on it, and rein their headlong speed,
> And bow their burning crests, and glide in fire
> Under the waters of the Earth again.
>
> (70–82)

The fauns are mystified by the nature of the 'spirits' which play so prominent a part in the act. Yet their bewilderment acts as a spur to celebration of a cycle which is at once natural and imaginative, and removes the film of familiarity from the ordinary. Here, creation of a fiction encourages the reader to see with fresh eyes. Caesurae and internal rhyme are the smoothest of reins on the 'headlong speed' of Shelley's imagination. And the final word, 'again', implies a dance-like energy which could give rise to other inventions. As the First Faun elegantly inquires, 'If such live thus, have others other lives . . .?' (83).

This highly self-conscious scene points to the challenges and rewards offered by Shelley's poem as it moves towards the overthrow of Jupiter. For the most part, the prolific inventiveness of the writing does not lead to self-indulgent fantasy. In the central part of II. iv, the dialogue between Asia and Demogorgon, Shelley's language is gnomic, taxing, unornamented, and fascinating:

DEMOGORGON
>   —If the Abysm
> Could vomit forth its secrets:—but a voice
> Is wanting, the deep truth is imageless;
> For what would it avail to bid thee gaze

On the revolving world? what to bid speak
Fate, Time, Occasion, Chance and Change? To these
All things are subject but eternal Love.

ASIA

So much I asked before, and my heart gave
The response thou hast given; and of such truths
Each to itself must be the oracle.—

(114–23)

This exchange takes us to the core of the drama, and succeeds in
sustaining a twofold vision: scepticism about the possibility of
solutions to major problems of human existence, yet faith in 'eternal
Love', an abstraction that the entire act has striven to incarnate in
its fictions. Asia realizes the play's central insight: we are at once
free and fated to interpret. And in this sense the passage strikes a
cautious as well as exultant note: 'a voice | Is wanting' at the heart
of Shelley's lyrical drama. The 'voice' of hope and aspiration does
not exist beyond the individual will to show an affirming flame and
see reality in an optimistic light.

   Such a flame burns most fervently in the jubilant fourth act whose
visions express the poet's will to resolve the lyrical drama's underlying
tensions and contradictions. Thomas Weiskel takes Demogorgon's
injunction 'to hope, till Hope creates | From its own wreck the thing
it contemplates' (IV. 573–4) as showing that 'Like Stevens, Shelley
does not ultimately surrender desire, and he is forced into an ironic
doctrine of fiction in order to sustain it';[32] this view is given support
by the abruptness with which Shelley modulates out of the difficult,
cryptic dialogue between Asia and Demogorgon into one of the
poem's greatest fictions:

DEMOGORGON

Behold!

ASIA

The rocks are cloven, and through the purple night
I see Cars drawn by rainbow-winged steeds
Which trample the dim winds—in each there stands
A wild-eyed charioteer, urging their flight.

---

[32] *The Romantic Sublime: Studies in the Structure and Psychology of Transcendence* (Baltimore and London, 1976), 157.

> Some look behind, as fiends pursued them there
> And yet I see no shapes but the keen stars:
> Others with burning eyes lean forth, and drink
> With eager lips the wind of their own speed
> As if the thing they loved fled on before,
> And now—even now they clasped it; their bright locks
> Stream like a comet's flashing hair; they all
> Sweep onward.—
>
> (128-40)

Certainly, this vision of the chariots which convey to Asia historical moments ('the immortal Hours | Of whom thou didst demand' (140-1)) demonstrates the 'imaginative complexity' which Timothy Webb sees as the major difference between *Prometheus Unbound* and *Queen Mab*: 'In *Prometheus Unbound* the implications are explored with a more satisfying imaginative complexity but the central motivation is essentially that of *Queen Mab* as formulated in the Lucretian epigraph'.[33] In *Prometheus Unbound*, however, Shelley's imagination profoundly remodels the didactic impulse to 'free men's minds from the crippling bonds of superstitions'.[34] In her response to Demogorgon's 'Behold!', Asia achieves one of the richest triumphs over the visual in these first two acts. Here the swiftness with which Shelley frames and admits imaginative caprice—'And yet I see no shapes but the keen stars'—forces the reader to redefine the promise of visual description. Asia's is a vision of the possibilities offered by history and of the creative involvement by the individual in the formulation of those possibilities. The passage's rhythms are an apt match for the inventive pursuit of desire which is both the play's special strength and makes Act II a self-sustaining though not self-indulgent creation.

ACT III

'A voice to be accomplished' (III. iii. 67): Prometheus' phrase might serve to suggest Shelley's search for an adequate way of presenting the changed universe of Act III. In this act he relies heavily on an

---

[33] 'Shelley and the Religion of Joy', *SIR* xv (1976), 367.
[34] Trans. of Lucretian epigraph to *Queen Mab*, 'Shelley and the Religion of Joy', 367.

intricate system of echoes and 'mutually inflecting . . . symbols', in Hildebrand's phrase, both to validate the poem's intimations of perfection and to resist the pull towards wordlessness which such perfection entails. The reliance is at times too evident, but the pressures registered by the poetry contribute to its most achieved moments.

In Wasserman's view, 'It is the characteristic of Shelley's imagination that he conceived many of his poems in cosmic terms, postulating in them an entire universe, together with its spiritual physics.'[35] The question that his formulation prompts in relation to *Prometheus Unbound* is this: do Shelley's 'spiritual physics' begin by standing in for, but end up displacing, 'beautiful idealisms of moral excellence'? The act deploys images of 'exhalation' and 'music' to suggest states of renewal in a way that is at times over-fluent, too removed from the human:

IONE

Thou most desired Hour, more loved and lovely
Than all thy sisters, this is the mystic shell;
See the pale azure fading into silver,
Lining it with a soft yet glowing light.
Looks it not like lulled music sleeping there? . . .

PROMETHEUS

. . . And as thy chariot cleaves the kindling air,
Thou breathe into the many-folded Shell,
Loosening its mighty music; it shall be
As thunder mingled with clear echoes.

(III. iii. 69–73, 79–82)

Flowing and finely cadenced as the writing is, its focus on a rarefied detail of the drama's symbolic mechanism seems misplaced. The poetry delights in echo and anticipation, sending the reader back to Ione's vision of the Spirits of the Human Mind (I. 759–62) and forward to the 'Æolian modulations' (IV. 188) which Panthea hears in the final act. And yet Shelley's descriptive detail is not doing very much, for all the busyness with which it glances before and after. That the shell should be 'mystic' is loosely grandiloquent in one sense; in another, it is a way of licensing the poet's sense of the shell's place

---

[35] Wasserman, 328.

in an elaborate system of self-reference. It is to this system that the finicky 'soft yet glowing light' belongs.

Wordsworth's *Prospectus* to *The Excursion* outlines the humanist vision that a millennial poetry should not neglect; after declaring that the 'Mind of Man' (40) is his song's 'haunt' (41) and 'main region' (41), Wordsworth goes on to proclaim that 'Beauty—a living Presence of the earth, | Surpassing the most fair ideal Forms | Which craft of delicate Spirits hath composed | From earth's materials—waits upon my steps' (42–5). It is the apparent absence of 'Beauty' as a 'living Presence' that has alienated readers of *Prometheus Unbound*. However, where Shelley's fictionalizing 'craft' and the claims of human experience are united in this act, he writes with authority.

The opening scene of the act introduces Jupiter who is a far more sophisticated creation than his counterparts in *Queen Mab* or *The Revolt of Islam*. The sense that evil is self-imposed, a state of delusion, makes itself apparent in the scene's verbal strategy. Cleverly related to previous assertions in the drama, Jupiter's words are both resonant and hollow. His rhetoric backfires, exhausting itself in the moment of utterance:

> And though my curses through the pendulous air
> Like snow on herbless peaks, fall flake by flake
> And cling to it—though under my wrath's night
> It climbs the crags of life, step after step,
> Which wound it, as ice wounds unsandalled feet,
> It yet remains supreme o'er misery,
> Aspiring . . . unrepressed; yet soon to fall . . .
>
> (11–17)

Even before his overthrow, Jupiter is impotent to control 'It', the soul of man. Not only does his simile in lines 12–13 recall the similar image used in Act II to evoke the inevitable onset of revolution (II. iii. 36–42), it also vitalizes the heroic struggle which Jupiter wishes to expose as fantasy. Elsewhere, as in lines 25–33, Jupiter's words anticipate the language of subsequent scenes, suggesting that the character is a medium through which Shelley asserts symbolic significance. That the same images and symbols, reinterpreted, will support an opposing vision shows how central to the poem is Shelley's emphasis on the importance of perception.

Participation in a network of echo and anticipation is both present in and absent from the act's second scene. So far as the scene creates a 'psychic space'[36] that erases the memory of an unregenerate actuality, it is absent. Yet it is the peculiar beauty of the scene to achieve this erasure of actuality by degrees. The sonorous idiom of the opening is appropriate to the tyranny whose overthrow is being described, as Apollo's simile for the 'terrors of his [Jupiter's] eye' (4) reveals: 'Like the last glare of day's red agony | Which from a rent among the fiery clouds | Burns far along the tempest-wrinkled Deep' (7–9).

It is in Ocean's fine speech, however, that Shelley offers a convincing evocation of an altered 'henceforth':

> Henceforth the fields of Heaven-reflecting sea
> Which are my realm, will heave, unstain'd with blood
> Beneath the uplifting winds—like plains of corn
> Swayed by the summer air; my streams will flow
> Round many-peopled continents and round
> Fortunate isles; and from their glassy thrones
> Blue Proteus and his humid Nymphs shall mark
> The shadow of fair ships, as mortals see
> The floating bark of the light-laden moon
> With that white star, its sightless pilot's crest,
> Borne down the rapid sunset's ebbing sea;
> Tracking their path no more by blood and groans;
> And desolation, and the mingled voice
> Of slavery and command—but by the light
> Of wave-reflected flowers, and floating odours,
> And music soft, and mild, free, gentle voices,
> That sweetest music, such as spirits love.

(18–34)

The presence of 'Blue Proteus' and the 'Fortunate isles' gives solidity to the renewed, self-sustaining universe which the poetry imagines, and the force of Shelley's wording strengthens the reader's grasp of this changed state. In the second line, for instance, 'heave' is given sharpened impact by the comparison between the sea and 'plains of corn | Swayed by the summer air'. More excitingly, Shelley makes the physical daringly insubstantial, intermingling the literal and figurative in the comparison between Proteus' view of the sea's surface and the mortal view of the evening sky. The speech concludes with a contrast between 'the mingled voice | Of slavery and command'

---

[36] The phrase is B. Weaver's, Zillman, 517.

and 'That sweetest music, such as spirits love', reinforcing the drama's concern with 'voice' and 'music'. Lyrical, almost autonomous, the speech creates a credible vision of a redeemed world.

Yet Ocean's next speech achieves a richer permanence:

> The loud Deep calls me home even now, to feed it
> With azure calm out of the emerald urns
> Which stand forever full beside my throne.
> Behold the Nereids under the green sea,
> Their wavering limbs borne on the windlike stream,
> Their white arms lifted o'er their streaming hair
> With garlands pied and starry sea-flower crowns,
> Hastening to grace their mighty Sister's joy.
>   [*A sound of waves is heard.*]
> It is the unpastured Sea hung'ring for Calm.
> Peace, Monster—I come now! Farewell.
>
>   (41–50)

'Quite simply the finest Shelley ever wrote':[37] there is a poised serenity about these lines which persuades one to respect Bloom's judgement. The rhythms, in Coleridge's words about a couplet from *Venus and Adonis*, give 'the liveliest image of succession with the feeling of simultaneousness';[38] each line is self-sufficient yet moves easily into the next. 'Feed' anchors the reader in the real world rather than taunting one with memories of that world. But this world can be transformed. When Ocean asserts that he will 'feed' the 'loud Deep' with 'azure calm', his phrasing is as non-physical as the earlier 'aerial kisses' on which the Fourth Spirit's poet 'feeds' (I. 741), yet it comes across as the reverse of ethereal. The abstraction, 'calm', finds a visionary emblem in the lines which follow 'Behold'. It is a sign of the writing's 'calm' that Shelley satisfies visual expectations here. What turns the poetry from achievement into excellence is the admission of an as yet unsatisfied 'hung'ring for Calm'. 'Peace, Monster' invites the reader not to worry about 'the unpastured Sea'. But line 49 gives compact and memorable expression to that state of as yet not wholly appeased desire for harmony which lies at the heart of some of Shelley's finest poetry.

---

[37] *SM* 133.
[38] *Biographia Literaria: or Biographical Sketches of my Literary Life and Opinions*, ed. George Watson (1956; repr. with additions and corrections, 1965; London, 1971), 179.

Lack of compactness is the feature that critics have regretted in the two scenes which conclude the act. A nineteenth-century reviewer wrote:

the actual unbinding of Prometheus, the business of the drama, is so drawn out and so commented on that one hardly knows where it is. His immense long speech after his emancipation . . . is quite revolting . . . [and] the lover-like stuff is . . . a disgusting travestie of the kind, innocent, virgin pity in the attendant nymphs of the real Prometheus.[39]

The actual unbinding is brevity itself, but the drift of the reviewer's comments has relevance to a reading experience of the poetry. In Prometheus' long speech in III. iii. (6–63), Shelley tries to move beyond self-validating symbolism into statement about the ways in which his hero will use his freedom. Shelleyan love and liberation are on imaginative trial here. What spoils the speech is its inability— in the first half—to transform its symbolic idiom. Unfortunately, Prometheus' 'Cave' (10) seems to offer merely an indulgent escape:

> A simple dwelling, which shall be our own,
> Where we will sit and talk of time and change
> As the world ebbs and flows, ourselves unchanged—
> What can hide man from Mutability?—
>
> (22–5)

In line 25 Prometheus' non-human perspective makes for a disquieting detachment. However, Shelley's verse does not convince the reader that Prometheus' proposal to make 'Strange combinations out of common things' (32) constitutes a satisfying enterprise. Only in its account of 'lovely apparitions' (49) does the writing begin to shrug off the limitations which it has previously imposed on itself:

> And lovely apparitions dim at first
> Then radiant—as the mind, arising bright
> From the embrace of beauty (whence the forms
> Of which these are the phantoms) casts on them
> The gathered rays which are reality—
> Shall visit us, the progeny immortal
> Of Painting, Sculpture and rapt Poesy
> And arts, though unimagined, yet to be.
>
> (49–56)

---

[39] *Gentleman's Magazine*, 1848, Zillman, 521.

There can be no questioning of the marshalling intelligence at work here. Nevertheless, the passage does not break free from its roots in Platonism as Earth's vision of a shadow world did in Act. I. The thinner imaginative life is only partially compensated for by the self-inwoven syntax. Shelley avoids phantasmagoria even as he subsumes politics. His language is chastened into elegant abstraction by his respect for the great commonplaces of idealism. As authoritative a note as may be found in the passage occurs in that last line's gesture towards something still to be imagined, though later in the speech the phrase 'swift shapes and sounds' (60) has the unspecificity of the 'beautiful idealism' at its most unchallenging.

'Any representation of an ideal state is bound to seem unrealistic and too easily attained':[40] Peter Butter's comments on lines 130–4 of III. iv. could well introduce a discussion of the entire scene. For the most part, Shelley's language neither overstates its optimistic case nor underestimates the degree to which its vision is necessarily a matter of aspiration. None the less, though it is far subtler, the first half of the Spirit of the Earth's speech (33–85 in all) lacks the direct impact of this anticipation of change in *Queen Mab*:

> How sweet a scene will earth become!
> Of purest spirits, a pure dwelling-place,
> Symphonious with the planetary spheres;
> When man, with changeless Nature coalescing,
> Will undertake regeneration's work . . .
>
> (vi. 39–43)

There is a sturdiness about 'regeneration's work' (even if it is the sturdiness of revolutionary cliché) which the reader only finds intermittently in the third act. But the second half of the Spirit of the Earth's speech reveals the lyrical drama's compensating sophistication:

> I hid myself
> Within a fountain in the public square
> Where I lay like the reflex of the moon
> Seen in a wave under green leaves—and soon
> Those ugly human shapes and visages
> Of which I spoke as having wrought me pain,

---

[40] *'Alastor' and Other Poems; 'Prometheus Unbound' with Other Poems; 'Adonais'*, ed. P. H. Butter (London and Glasgow, 1970), 302.

Past floating through the air, and fading still
Into the winds that scattered them; and those
From whom they past seemed mild and lovely forms
After some foul disguise had fallen—and all
Were somewhat changed . . .

(61-71)

There is a fine restraint here (though the rhyme at the end of the
third and fourth lines seems lax), especially in the concluding words,
which communicate nonchalant delight that a state so often desired is
within reach. However, the vision of unmasking, prepared for by the
previous account of 'such foul masks with which ill thoughts | Hide
that fair being whom we spirits call man' (44-5), is not without a
certain utopian chill. Shelley exploits the gap between these 'mild and
lovely forms' and the reader's sense of the human condition to emphasize the fact that this is a poetic vision, the product of aspiration.

In the 'clear eloquence'[41] of its analysis of pre- and post-
revolutionary humanity, the Spirit of the Hour's speech (98-204)
surpasses anything in *Queen Mab*. Throughout the speech Shelley's
figurative language sharpens rather than devitalizes the force of his
argument:

             the impalpable thin air
And the all-circling sunlight were transformed
As if the sense of love dissolved in them
Had folded itself round the sphered world.
My vision then grew clear, and I could see
Into the mysteries of the Universe.

(100-5)

'Spiritual physics' work lucidly here, adequately imaging transformation by refusing to be wholly figurative: 'the sense of love',
the reader is persuaded to feel, has actually 'dissolved' into the 'air'
and 'sunlight'. Interestingly, the speech is most impressive when
expressing its fascination with 'fallen' states of mind and soul:

None frowned, none trembled, none with eager fear
Gazed on another's eye of cold command
Until the subject of a tyrant's will
Became, worse fate, the abject of his own . . .

(137-40)

[41] *SM* 138.

> None talked that common, false, cold, hollow talk
> Which makes the heart deny the *yes* it breathes
> Yet question that unmeant hypocrisy
> With such a self-mistrust as has no name.
>
> (149–52)

In the second extract, 'as has no name' concedes too quickly the struggle to find an appropriate way of talking about 'self-mistrust'. But line 151 shows a remarkable sympathy with 'unmeant hypocrisy' (an insight probably indebted to but deeper than Coleridge's 'words of unmeant bitterness'[42]).

In its fascination with 'The ghosts of a no more remembered fame' (169) that it tries to dismiss, the final paragraph of blank verse reveals the pressures which build up inside a poetry devoted to millennial change. The language raises the ghosts which it seeks to exorcize. The 'foul shapes, abhorred by God and man' (180) referred to by Shelley assume, through the pressure of echoes, an odd power. If the 'foul shapes' oppose the 'shapes' which Shelley would pit against them, they also remind the reader of the transience in which the entire work's 'shaping' is involved. After all, they 'Were Jupiter' (183). As the passage progresses, the perfection that Shelley envisages excites his imagination. An illustration is supplied by the sweeping rhetorical momentum of these lines:

> The painted veil, by those who were, called life,
> Which mimicked, as with colours idly spread,
> All men believed and hoped, is torn aside—
>
> (190–2)

The desire to change actuality has led to a wish to do away with the life lived by 'those who were', and the reader may be surprised by Shelley's phrasing in 'All men believed and hoped', especially if he or she remembers Earth's more generous reference to 'all that faith creates, or love desires' (I. 201). But Shelley's awareness of his will to repudiate is apparent in lines concluding an act which is, for all its apparent absence of dramatic conflict, often rewardingly at odds with itself:

> Nor yet exempt, though ruling them like slaves,
> From chance and death and mutability,
> The clogs of that which else might oversoar

[42] *Christabel*, 665.

> The loftiest star of unascended Heaven
> Pinnacled dim in the intense inane.

(200-4)

'Unascended' subverts the achievement glimpsed by 'Pinnacled', even though the lines are cadenced as though celebrating a triumph. The tension reflects an idealism which both ignores and registers limits.

ACT IV

When Shelley writes in the final act—

> Language is a perpetual Orphic song,
>   Which rules with Dædal harmony a throng
> Of thoughts and forms, which else senseless and shapeless were.

(415-17)

—he suggests a lofty disregard for the 'throng | Of thoughts and forms' which 'Language' retrieves from 'shapeless' chaos. 'Orphic song', as has been suggested already, results at times in too facile an exploitation of a self-enfolding symbolic idiom; it can enchant but ignore the mute particulars of experience. Yet at its best, both in the fourth act and in the entire poem, Shelley's language is sensitized to the individual nature of 'thoughts and forms' that it seeks to 'rule'.

Act IV often behaves as if it can take for granted that the ideological battle has been won. In a sense, its emphasis on 'harmony' does not so much evade the referential as imply delight that language can correspond faithfully to the poet's vision of a redeemed world. This is most clearly illustrated by the great blank verse of the fourth act, the 'visions' of Ione and Panthea. Here, as in the finest passages in *Epipsychidion*, self-consciousness is a prelude to compelling writing:

IONE

>   Even whilst we speak
> New notes arise . . . What is that awful sound?

PANTHEA

> 'Tis the deep music of the rolling world,
> Kindling within the strings of the waved air
> Æolian modulations.

IONE

Listen too,
How every pause is filled with under-notes,
Clear, silver, icy, keen, awakening tones
Which pierce the sense and live within the soul
As the sharp stars pierce Winter's chrystal air
And gaze upon themselves within the sea.

(184–93)

The poet suggests both the visionary timbre of his words and in the line—'Which pierce the sense and live within the soul'—their ambitious design. Shelley sets himself to listen to those 'under-notes' with the same reverence that an earlier poet would have listened to the music of the spheres. In Ione's vision, the impulse to celebrate remains, but the content of her celebration assumes a more autonomous character:

I see a chariot like that thinnest boat
In which the Mother of the Months is borne
By ebbing light into her western cave
When she upsprings from interlunar dreams,
O'er which is curved an orblike canopy
Of gentle darkness, and the hills and woods
Distinctly seen through that dusk aery veil
Regard like shapes in an enchanter's glass;
Its wheels are solid clouds, azure and gold,
Such as the genii of the thunderstorm
Pile on the floor of the illumined sea
When the Sun rushes under it; they roll
And move and grow as with an inward wind.

(206–18)

This description is written with a restraint that half-incites and half-enacts wariness about its referential content. Shelley is most wary in the way his phrase, 'as with an inward wind', evades the literal status of its possible source in *Paradise Lost*,[43] where Christ's chariot is said to be 'Itself instinct with spirit' (vi. 752). The poetry deals in simile. Ione describes a chariot 'like that thinnest boat'; the

---

[43] The suggestion that Ione's vision attempts to ''transume' 'the Miltonic chariot [of Paternal Deity]' is Harold Bloom's, *Poetry and Repression: Revisionism from Blake to Stevens* (New Haven and London, 1976), 94. The connection between Milton's chariot and Panthea's vision has been made by A. Oras, Zillman, 585–6.

canopy's 'dusk aery veil' makes hills and woods 'Regard like shapes in an enchanter's glass'; the chariot's wheels 'roll | And move and grow as with an inward wind'. Simile is Shelley's way of preserving the uniqueness of his subject, yet bringing it within the reach of the reader's imagination. Ione's lines are at once assured and reticent about the myth-making in which they are engaged; for instance, the speech riddlingly yokes 'Distinctly seen' with 'that dusk aery veil', authenticating the reality of the vision by thwarting the reader's visualizing instincts. This reality is sharpened by a subtle use of negation when Shelley writes of 'Tempering the cold and radiant air around | With fire that is not brightness' (229–30). To gloss these lines as a reference to Herschel's 'dark rays'—'infrared emanations that produce heat without light—which, Davy suggested, might be given off by the moon' (PP 201)—is helpful only if the extent to which Shelley's language transforms any supposed 'subject' is kept in mind. That said, Shelley's enthusiasm for scientific discoveries is, as various commentators have remarked, perceptible in the details of both visions. Cronin argues that the 'scientific and the visionary modes of apprehension cannot be irreconcilably opposed for both have the same origin, both have their home in the mind of man'.[44] But the language places its own workings so firmly in the foreground that the reader is more struck by the creation of new verbal 'relations'—as in the flickering paradox, 'cold and radiant'—than by an external reality to which the words may refer. Throughout, the language has an inward-looking, unassertive intelligence.

Panthea's vision grapples more demandingly with the exhilaration of its own energy:

> And from the other opening in the wood
> Rushes with loud and whirlwind harmony
> A sphere, which is as many thousand spheres,
> Solid as chrystal, yet through all its mass
> Flow, as through empty space, music and light:
> Ten thousand orbs involving and involved,
> Purple and azure, white and green and golden,
> Sphere within sphere, and every space between
> Peopled with unimaginable shapes
> Such as ghosts dream dwell in the lampless deep
> Yet each intertranspicuous, and they whirl

---

[44] Cronin, 166.

Over each other with a thousand motions
Upon a thousand sightless axles spinning
And with the force of self-destroying swiftness,
Intensely, slowly, solemnly roll on—
Kindling with mingled sounds, and many tones,
Intelligible words and music wild.—

(236–52)

Ignoring the usual signals of vision in *Prometheus Unbound*—'behold'
or 'see'—Shelley imitates the formidable start of his Miltonic original.
Contrast reveals a more urbane and supple use of language in the
lyrical drama: in Milton we read: 'forth rushed with whirlwind
sound | The chariot of Paternal Deity' (*Paradise Lost*, vi. 749–50),
whereas Shelley discovers a 'loud and whirlwind harmony'. When
Harold Bloom sees the Romantic poet as relying on 'speed' to
'surmount the peculiar strength of [Milton's] anterior vision' of
the angelic dance in Book V (618–27) of *Paradise Lost*,[45] he
responds too hastily to 'the force of self-destroying swiftness', a
line which provokes an unexpected reply: 'Intensely, slowly, solemnly
roll on'. The passage takes pleasure in qualifications which mute
even if they do not impede its onward momentum. In lines 241–3,
for instance, the lines are rounded out, exhibiting what D. J. Hughes
(with lines 243–6 in mind) calls the 'sufficient plenitude'[46] of the
vision. And in lines 243–6, Shelley has time for a transmuted
Gothicism that is gravely balanced between imagination and caprice.
But 'caprice' is an ungenerous description of the quick decision
in 'intertranspicuous' to contest the unimaginableness of these
'shapes'—not by describing them, but by relating them to something else (the objects of ghostly dreaming) so that their specific
nature seems of no relevance. Kenneth Cameron cogently defends
a scientific interpretation of this passage: 'Shelley's picture, then,
is a kind of "dance of matter," with the compounds, elements,
molecules, and atoms depicted as orbs whirling within orbs, the
electric and other "fluids" depicted as interpenetrative shapes flowing
between them.'[47] But, again, what impresses in the poetry is an
inventiveness which reclaims scientific theory for the visionary
imagination.

[45] *Poetry and Repression*, 97.
[46] 'Potentiality in *Prometheus Unbound*', 147.
[47] Cameron, 550.

The poetry's transformation of reality into its own visionary image and likeness is exciting, but gives more urgency to the complicated issue of meaning. D. J. Hughes would collapse the distinction between language, however self-conscious, and subject-matter, however elusive: 'The passage [IV. 180–318] seems to me, as does all of *Prometheus Unbound*, an attempt to mirror the creative mind at full stretch, and ultimately to be self-reflexive, about itself.'[48] This sentence contains a crucial shift of emphasis. For a poem to 'mirror the creative mind at full stretch' is one thing; for it 'to be self-reflexive, about itself' is another. Shelley glimpses reflections of his own poetic activity in the mirror of his 'multitudinous Orb' (253), but the relationship, however real, is oblique and cannot be reduced to a simple equation between the 'Orb' and poetry. Panthea's vision seeks to dematerialize the world of matter and recompose it as a purely linguistic energy. Yet its authority lies in the resistance which the materials of the vision offer to the language, even as they are being 'kneaded into one aerial mass | Which drowns the sense':

> With mighty whirl the multitudinous Orb
> Grinds the bright brook into an azure mist
> Of elemental subtlety, like light,
> And the wild odour of the forest flowers,
> The music of the living grass and air,
> The emerald light of leaf-entangled beams
> Round its intense, yet self-conflicting speed,
> Seem kneaded into one aerial mass
> Which drowns the sense.

> (253–61)

It is important to see that these lines dispute the impulse to describe their own activity. The 'self-conflict' which validates yet complicates the writing's 'speed' is fought between poetry's consciousness of itself and of its subject. And it is while this friction induces both suspension and imminence in the verse that Shelley writes great poetry. When, in the last few lines, the conflict resolves itself into a visionary triumph—signalled by the sharp internal chime of 'speed' and 'kneaded'—over 'the sense' and over matter (now recomposed as 'one aerial mass'), there is a distinct feeling of reaching a limit as well as achieving a victory. That the climax of a vision restores that

---

[48] 'Potentiality in *Prometheus Unbound*', 158.

finiteness which it insists that it overcomes could have been a sorrowful Romantic irony. Yet Shelley survives his own deflation (Ione's explanation of the Spirit of the Earth's smile as 'only mocking the Orb's harmony' (269)), and launches into a more obviously referential account of 'the melancholy ruins | Of cancelled cycles ' (288–9). Radical geology involves harder, more assertive poetry. Shelley glosses the beams shot from the infant's forehead as 'Embleming Heaven and Earth united now' (273), and, probably influenced by a passage in Keats's *Endymion*,[49] he wins power by surrendering to the objectness of his materials: 'prodigious shapes | Huddled in grey annihilation, split, | Jammed in the hard black deep' (300–2). Yet it is not merely the sharp-edged physicality of the poetry that explains its force. Those 'prodigious shapes', so much vaguer than the Keatsian details[50] of 'anchors, beaks of ships, | Planks turned to marble, quivers, helms and spears' (289–90), have their own precision (evident in the sure rhythms). Locating an indefiniteness, the poet creates imaginative room for himself.

The final lines of the speech interiorize the 'cancelled cycles' in the poem's structure as they witness their own collapse:

> till the blue globe
> Wrapt Deluge round it like a cloak, and they
> Yelled, gaspt and were abolished; or some God
> Whose throne was in a Comet, past, and cried—
> 'Be not!'—and like my words they were no more.
>
> (314–18)

These lines have been finely discussed by D. J. Hughes who writes: 'The command "Be not!", which can be taken as the poetic motivation of the entire scene (180–318), cancels not only the phenomenal world, but also the Word that Shelley has been building throughout.'[51] Shelley's capacity to build and abolish his poetic Word attests both to the poem's sense of its power—its capacity to imagine transformation—and its awareness of its limits—such a transformation is only taking place within a poem. He creates an effect of cancellation on two occasions in the two parts of the vision: the first is a loss of 'sense', drowned by its own capacity for synaesthesia; the second is a separateness induced by the spectacle of things which

---

49 iii. 119–36. See Shelley's comments on *Endymion*, *Letters*, ii. 252.
50 See *Endymion*, iii. 123–4.
51 'Potentiality in *Prometheus Unbound*', 150.

were 'mortal but not human' (298), a separateness which suddenly converts itself into the separate God-like presence of the poet who can abolish the phenomenal and even that system of reference on which he seems to lean (not 'God', but 'some God' is invoked).

The ease with which the drama restates its ideological position at the conclusion is striking. Yet, though Demogorgon's coda situates itself beyond self-questioning, it stresses the need for constant re-creation of reality, the need to 'hope, till Hope creates | From its own wreck the thing it contemplates'. The poetry strives to end up with a glimpse of language as unfrail spell, talismanic charm. So the dramatis personae say to Demogorgon: 'Speak—thy strong words may never pass away' (553). But *Prometheus Unbound* is memorable precisely because the fear that its words may 'pass away' has, throughout, prompted the inventiveness of its language.

6

The Witch of Atlas
Fictions, Visionary Rhyme,
and Human Interest

Like *The Sensitive-Plant*, *The Witch of Atlas* pushes to an extreme
that impulse to create fictions which is prominent in *Prometheus
Unbound*. 'Fictions' is an especially useful word to describe what
goes on in *The Sensitive-Plant* and *The Witch of Atlas* because it
suggests not only the imagination's products but also the process
which results in these products (as in the phrase 'a work of
fiction').[1] Moreover, in both poems, 'fictions' often betray or assert
in various ways an awareness of their fictional status, swinging
between the seeming confession that they are 'only' fictions, made-
up, fabricated, and the implicit claim that their kind of feigning
offers a more valuable experience than didactic or realistic alterna-
tives. In *The Sensitive-Plant* and *The Witch of Atlas* Shelley's
management of this awareness is subtle and understated. Radically
self-knowing (as opposed to simply knowing), both poems seem
to impose limits on their degree of seriousness—then, at crucial
moments, to transgress those limits. *The Sensitive-Plant* alters
the reader's grasp of what seriousness in poetry is when, in its
Conclusion, it offers, with almost shoulder-shrugging casualness,
a 'modest creed' (13) to set against the fable's pessimistic drift.
Switching tones, though, Shelley goes on to urge the claims of this
idealist 'creed' with eloquent force: the merest of fictions speaks with
the accents of ultimate truth:

> That garden sweet, that lady fair
> And all sweet shapes and odours there
> In truth have never past away—
> 'Tis we, 'tis ours, are changed—not they.

---

[1] This distinction is drawn from Victor Sage's helpful essay on 'Fiction', Fowler
(ed.), *A Dictionary of Modern Critical Terms*, 71–4, esp. 72.

For love, and beauty, and delight
There is no death nor change: their might
Exceeds our organs—which endure
No light—being themselves obscure.

(Conclusion, 17–24)

Yet hope is grounded in uncertainty, in the poet's sceptical distrust of our 'organs'. It grows out of 'vacancy', to use the word Shelley employs in various contexts, most notably the conclusion of 'Mont Blanc', for that emptied state of mind and sense of reality which is one possible response to the awareness that meaning is constructed by the 'human mind's imaginings'. Shelley may appear to have settled for a Spenserian sureness about what reality 'really' is, to refer back to the Renaissance poet's account in *The Faerie Queene* of the Garden of Adonis where decay is illusory since 'The substance is not chaunged, nor altered, | But th'only forme and outward fashion' (III. vi. 38). But the lines quoted from the Conclusion are touched with a pathos which is latent in Spenser's admission that 'formes are variable and decay' (III. vi. 38). This 'pathos' has to do with, and is suggested by, the way the poetry behaves, its air of holding sadness at bay. What matters here is Shelley's presence (or the illusion which he creates of 'presence'), the workings of his speaking voice. The conversational emphases—'In truth', ' 'Tis we, 'tis ours'— prepare the ground for the assertions of the last stanza. Again, his voice dips significantly in the final line and a half. Reasserting the inadequacy of our 'organs', Shelley explains his 'creed' as a reaction against unpalatable sense-impressions. It is as if he were trying to console himself and his reader for their involvement with the 'garden sweet' and 'lady fair', offering an affirmation which can be taken either as 'merely fictional' or as a 'Supreme Fiction'. By a variety of methods—principally its manipulations of perspectives and figurative language—the entire poem has put awareness of itself as a fiction in the foreground. Yet it does not sacrifice emotional impact to self-consciousness; it engages the reader's feelings by laying bare how any vision of existence is bound up with fictionalizing.

The Witch of Atlas shares this capacity to stay within, but move beyond, tonal limits, to deepen into perceptions which have emotional power, yet from which the poem can stand at a remove. The famous transition from stanza LXII to stanza LXIII provides an example:

'This,' said the wizard maiden, 'is the strife
Which stirs the liquid surface of man's life.'
And little did the sight disturb her soul—
    We, the weak mariners of that wide lake
Where'er its shores extend or billows roll,
    Our course unpiloted and starless make
O'er its wild surface to an unknown goal—
    But she in the calm depths her way could take
Where in bright bowers immortal forms abide
Beneath the weltering of the restless tide.

                                        (543–52)

It is writing such as this which substantiates the assertion made
by Reiman and Powers: 'The tone holds the reader's feelings in
suspension, as the poet describes the incomparable beauty and
perfection of the Witch and, at the same time, her lack of understanding sympathy with the problems of mortal creatures.' (PP 347.)
The achievement of these lines is to touch us by touching so lightly
on 'the common feelings of men' to which Mary Shelley robustly
wished her husband's poems had been 'more addressed' (PW 388).
The placid detachment of the Witch is registered amusingly at first;
her words trot out glibly; her rhymes point a moral which sounds
more than a shade self-satisfied. But the subsequent change of tone—
'And little did the sight disturb her soul'—is less indulgent; though
in no way reproving, it causes a chasm to yawn momentarily between
reader and character. However, Shelley handles his alienation effect
with restraint, delicacy. In the ensuing process of working out
his comparison between human beings and 'weak mariners' (a
comparison whose extended development gives it the self-engrossed
air of a conceit), Shelley just glances at a poignancy too sharp
to need underlining. At the same time, he derives and offers a
compensating pleasure from the working-out. The Witch of Atlas
moves with bewildering speed between alluding to our reality and
relishing its own imaginative autonomy.[2]
    Shelley's ways of communicating his presence in this poem,
as in The Sensitive-Plant, are highly sophisticated. Whether this

_____

[2] This to-and-fro movement is slighted by deconstructionist analyses of the poem
such as Jerrold Hogle's stimulating reading with its emphasis on 'metaphors and
metamorphoses which sooner or later must reveal their strictly transformative process',
'Metaphor and Metamorphosis in Shelley's "The Witch of Atlas"', 333.

sophistication depends upon something as deliberated as the
'Nabokovian aloofness from the conventions within which [Shelley]
is writing' that Richard Cronin discerns is arguable.³ What needs
to be said, though, is that the poem's involvement with its fictions
is its chief strength. As a series of fictions, *The Witch of Atlas* flirts
throughout with the suggestion of allegory to which Shelley invites
our attention at the end of the introductory stanzas:

> If you unveil my Witch, no Priest or Primate
> Can shrive you of that sin, if sin there be
> In love, when it becomes idolatry.

> (46–8)

Is the poem an allegory which uses its central fiction as a 'veil' to
conceal, yet arouse interest in, abstract ideas? If not, is it simply an
exercise in escapism? The issue is complicated and requires some
disentangling. One can, I think, answer both questions with a
qualified 'no': qualified because one needs to invoke the idea of
'reflexivity' to describe the way Shelley's imaginings work, yet one
needs, too, to be wary of using it merely as a fashionable substitute
for 'allegory' on the one hand, 'escapism' on the other.

Few critical manœuvres these days are more predictable than the
reading which tells us that a poem is 'reflexive' because it is 'about
itself', concerned with its own imaginative workings, processes of
signifying. But Keach directs us to the suggestion of the word which
is most relevant to my argument when he asserts that '[reflexive]
locutions call unusual attention to the act of mind they presuppose
in the writer and provoke in the reader'.⁴ It is such a summoning
of 'unusual attention' which Shelley's fictions perform in *The Witch
of Atlas* Yet their very self-sufficiency often prevents the reader from
regarding them as being 'about poetry' in any straightforward way.
The lines above 'provoke' the reader intriguingly; they both warn
and incite. Shelley pretends to scold our 'sin' (interpretative zeal;
desire for significance). But the last line and a half alter direction,
come alive with a sense of quickened wonder which, it is suggested,
we, too, shall experience. We may start as analysts; we shall end
up as worshippers. 'Idolatry' is precise and resonant. The word
recovers an 'innocent' meaning, that of 'image worship'. In doing
so it defines an appropriate response to a poem so inventive in

³ 'Shelley's Witch of Atlas', *KSJ* xxvi (1977), 90.   ⁴ Keach, 79.

creating images, fictions. As 'love' becomes 'idolatry', the reader participates in a process that mimes the poet's own Pygmalion-like discoveries—'Pygmalion-like' because Shelley is obsessed in his 'visionary rhyme' (8) with that 'queer business of using one's imaginative experience to create something surprising to oneself'.[5]

Yet the process of imagining and responding to imagining is not so uncomplicated, innocent, purely aesthetic. There are darker implications as well. Shelley implies that the poet's 'love', the capacity for 'identification . . . with the beautiful which exists in thought, action, or person, not [his] own' (A Defence of Poetry, PP 487), can become 'idolatry', fixation on his imaginative creations. And he suggests that the reader can experience a comparable narrowing of focus. Shelley dislikes concentration on one object because it may lead to neglect of infinite potentiality: 'Love is like understanding, that grows bright, | Gazing on many truths', he would write in Epipsychidion (162–3). That creation and love of the particular might lead to stasis and fixation is a fear which lies at the heart of The Witch of Atlas. It is at the heart, too, of Shelley's idealist poetic, a consequence of the belief that 'when composition begins, inspiration is already on the decline' (A Defence of Poetry, PP 504).[6] True, the dilemma is more a source of amusement than anxiety in the poem. But Shelley handles it with unusual awareness of the issues at stake. The zest and vitality of his fictions act as defences against the allegorical 'unveiling' which constitutes 'idolatry' in the darker sense I have been discussing.

Shelley asserts the superiority of his creation's creation, the Witch's Hermaphroditus, over the invention of Pygmalion in these lines from stanza XXXV, lines which mirror and evaluate the poet's own procedures:

> And a fair Shape out of her hands did flow—
> A living Image, which did far surpass
> In beauty that bright shape of vital stone
> Which drew the heart out of Pygmalion.
>
> (325–8)

---

[5] G. Wilson Knight, The Starlit Dome: Studies in the Poetry of Vision (1941; London, Oxford, and New York, 1971), 226.

[6] See the interesting discussion of Shelley's theory of poetry in Leighton, esp. pp. 44–7 where she argues that, for Shelley, 'Writing is . . . by its very nature a process of loss', p. 45.

The sinuous *ottava rima* and the flickering wit of the final rhyme ('stone'/'Pygmalion') recall, while holding at arm's length, a Byronic knowingness. Shelley knows that his is not the simpler dilemma of the artist torn between artifice and reality; rather, he is torn between two attitudes to poetic invention. There is no way of avoiding fictionalizing. The Witch 'surpasses' Pygmalion because she is free of that emotional need which the concluding cadences of stanza XXXV evoke strongly. The language, however, latches on to the ambiguities of such a superiority. The Witch's creation, the 'living Image', is set against Pygmalion's 'bright shape of vital stone' with teasing ironic intent. It is disturbingly lacking in the power to disturb which 'vital' possesses. If the Witch avoids Pygmalion's version of 'idolatry', she substitutes her own: delight in the reflexive activity of the imagination.

No impulse in Shelley's poetry is more unsettling for the reader who wishes to claim that it is about something than such reflexive delight. Donald Davie makes an acute perception which offers some comfort. Commenting on *The Witch of Atlas*, he says: 'Like "Alastor" and "The Sensitive Plant" it has no meaning except as a whole. It is one half of a vast metaphor with the human term left out; and this, its meaning for human life, emerges from the shape of the whole or else it is lost for ever.'[7] This statement at once stares into an abyss of figurative inventiveness so dizzying that all meaning is lost for ever and offers an escape. However, it is closer to a reading experience of the poem to say that its 'meaning for human life' glimmers, withdraws, and tantalizes. All depends on the varying degrees to which we are made to feel that a full account of the claims of human existence has been omitted. Shelley shapes the reader's awareness of such omissions by 'double takes' which involve us in seeing at the same time the human meaning of an experience or idea and the meaning the experience or idea possesses for the Witch. For instance, we respond to the neutralizing of the moral suggestions of 'love', elsewhere a key Shelleyan concept, in the following lines:

> Then by strange art she kneaded fire and snow
>   Together, tempering the repugnant mass
> With liquid love—all things together grow
> Through which the harmony of love can pass . . .

(321–4)

---

[7] Davie, 157.

The 'fair Shape' which emerges from this 'strange art' bears a relation
to 'the harmony of love' which is analogous to that of Shelley's
fictions to their imaginative source. 'The harmony of love' directs
our attention not so much to the laudable nature of the Witch's
activity as to the perpetual release of creative energy that is central
to her nature. For all the poem's inventiveness, there is a constant
if subliminal recognition that 'fictions' limit, even as they appease,
the craving of the imagination. The double use of 'love' in the above
lines could be regarded as signalling this recognition. In line 323,
'liquid love' is as much image as abstraction: it implies agencies of
design and form which inhere in the created product. In line 324,
'the harmony of love' is an imageless abstraction and suggests a
shaping spirit which precedes and transcends the product through
which it manifests itself.

Of course, there are many occasions in the poem when the craving
of the imagination is appeased, when the poem's current flows from
the imageless to the imaged. Such a flow is dramatized in stanza XX:

> And wondrous works of substances unknown,
>   To which the enchantment of her father's power
> Had changed those ragged blocks of savage stone,
>   Were heaped in the recesses of her bower;
> Carved lamps and chalices and phials which shone
>   In their own golden beams—each like a flower
> Out of whose depth a fire fly shakes his light
> Under a cypress in the starless night.
>
>   (201–8)

In the opening disclaimer of knowledge, 'wondrous works of
substances *unknown*' (my italics), Shelley admits and licenses the
need for imaginings. The ensuing lines dazzlingly reverse the initial
concession to the 'unknown'. The 'lamps and chalices and phials'
are self-containedly metaphoric, shining reflexively 'In their own
golden beams',[8] and transformed through simile by an imagination
constantly on the look-out for comparisons. Yet the poetry's effect,
though playful, goes beyond playfulness as the final comparison
works itself out. 'Starless' is a word which Shelley will later use as
an epithet for the human condition, 'Our course unpiloted and
starless' (548). Here it lends to Shelley's image a sombre shading at

---

[8] See the account of reflexive imagery in these lines in Keach, 112–13.

odds with the resource which the writing demonstrates. For
a moment, the creative spark is as tiny an illumination as
the fire fly's against the 'starless' night. And, for a moment,
the reader glimpses how the poem, in its commitment to
the imagination, displays a kind of courage.

So far as the Witch is concerned, 'Love', the poem's
presiding presence, enters into a more slippery liaison with
'morals' than the marriage between the two which *A Defence
of Poetry* would proclaim: 'The great secret of morals is
Love; or a going out of our own nature, and an identification
of ourselves with the beautiful which exists in thought,
action, or person, not our own. . . . The great instrument
of moral good is the imagination; and poetry administers
to the effect by acting upon the cause.' (*PP* 487–8.) Phrases
such as 'her love' suggest less the Witch's capacity to identify
with 'the beautiful which exists in thought, action, or
person, not our own' than that quality in her which incites
Shelley's imagination. Stanza XIII supplies an example:

> Which when the lady knew she took her spindle
>   And twined three threads of fleecy mist, and three
> Long lines of light such as the Dawn may kindle
>   The clouds and waves and mountains with, and she
> As many star-beams, ere their lamps could dwindle
>   In the belated moon, wound skilfully;
> And with these threads a subtle veil she wove—
> A shadow for the splendour of her love.

> (145–52)

In that last line, 'A shadow for the splendour of her love', 'her love'
seems interchangeable with, say, 'her essence'. The poetry returns to
it as to an originating source, yet in *The Witch of Atlas* it is the
excursions from pure source into language and fiction which hold our
interest. Here, the preceding lines make such an excursion with their
description of the 'veil' that the Witch needs to weave for herself as a
'shadow' to protect onlookers. The texturing of this veil is conveyed
through words which are themselves 'wound skilfully'. Syntax
proclaims its free and easy intentions when Shelley ends a clause with
a preposition, 'with'. And yet he manages to keep control over his
amusingly spun-out materials. The poetry seems on the verge of
stumbling, but maintains its balance with seemingly careless elegance.

It is the very unidealistic verve of the writing which distinguishes it from other accounts in Shelley's poetry of a veiled ideal. Even the rhyme of 'kindle' and 'dwindle' is unemphatic. Though it echoes a crucial rhyme in the 'Life of Life' lyric, it does not point, as it does in *Prometheus Unbound*, towards the difficulty of sustaining inspiration.[9] Instead, it enacts a witty parody of that theme. Shelley contrives to appear as a rhyme-juggling, improvising poet, living dangerously on his wits. In this stanza the rhyme prompts the delightfully particularized detail of the 'belated moon'. In *Prometheus Unbound* such a detail would require the reader to sense—to quote William Hildebrand again—'the shape of that intricate trellis of mutually inflecting metaphors and symbols that constitutes Shelley's ontological and epistemological intuitions in the drama'. Here, though, the detail does not point towards a more general, shadowy significance; less signpost than eye-catching find, it relishes its independence from some larger design. When Shelley describes the product of the Witch's weaving as a 'subtle veil', the word 'subtle' is both funny and justified—funny because it recalls and implicitly mocks the more straightfaced subtlety of works like *Prometheus Unbound*, justified because the poem's dealings with reflexive suggestions are so adroit. 'Subtlety' in *The Witch of Atlas* is the result of the poet's stance towards his material: quizzical, playful, detached, self-aware in ways which check impulses (his and ours) to allegorize. But while Shelley's playful tone disables over-ponderous allegorizings of his poem as being a poem about poetry, his fictions in *The Witch of Atlas* manage to communicate, as few other poems do, the mental processes from which they take their origin.

Worrying the issue of the imagination's relationship with morality in the poem, Leigh Hunt had this to say:

The Witch of Atlas . . . is but a personification of the imaginative faculty in its most airy abstractions; and yet the author cannot indulge himself long in that fairy region, without *dreaming* of mortal strife. If he is not in this world, he must have visions of it. If fiction is his reality by day, reality will be his fiction during his slumbers.[10]

Hunt begins with an equation between the poem and 'a personification

[9] Cf. *Prometheus Unbound*, II. v. 48 and 50. See also D. J. Hughes, 'Kindling and Dwindling: The Poetic Process in Shelley', *KSJ* xiii (1964), 13–28.
[10] From a critique of Shelley's *Posthumous Poems*, in Redpath, *The Young Romantics and Critical Opinion*, 409.

of the imaginative faculty' that is hard to avoid. Yet it testifies more to a critical desperation to impose meaning than to any identification of the Witch explicitly urged by the poet.[11] Hunt goes on to pay a tribute to his friend's moral zeal—'reality will be his fiction during his slumbers'—which conceals a more intriguing, perhaps more disquieting insight: that, ousting the moral from the centre of the poem, Shelley delights in showing how reality can be treated as another fiction. He is a sophisticated antagonist in *The Witch of Atlas* of that 'Nature' which Spenser invokes in his account of False Florimell, the 'Nature' which grudged to 'see the counterfet should shame | The thing it selfe' (*The Faerie Queene*, III. viii. 5.). In one sense, Shelley positively flaunts the 'counterfet' qualities of his poem; in another, he upgrades the status of its feigning or counterfeiting, presenting it as a 'visionary rhyme'. The context of Shelley's use of the word 'visionary', his attempt to coax Mary Shelley into accepting the validity of his enterprise, gives the term an ambivalent flavour. The poem is 'visionary' in a self-deprecatory sense, self-exiled from sturdy reality. But 'visionary' also carries a positive charge; it implies that *The Witch of Atlas* penetrates further into the nature of the 'real' than a stubbornly naturalistic work such as Wordsworth's *Peter Bell*—the butt of Shelley's airy, good-humoured contempt in the introductory stanzas:

> My Witch indeed is not so sweet a creature
>   As Ruth or Lucy, whom his graceful praise
> Clothes for our grandsons—but she matches Peter
>   Though he took nineteen years, and she three days
> In dressing. Light the vest of flowing metre
>   She wears: he, proud as dandy with his stays,
> Has hung upon his wiry limbs a dress
> Like King Lear's 'looped and windowed raggedness.'
>
> If you strip Peter, you will see a fellow
>   Scorched by Hell's hyperequatorial climate
> Into a kind of a sulphureous yellow,
>   A lean mark hardly fit to fling a rhyme at;
> In shape a Scaramouch, in hue Othello.
>
>   (33–45)

---

[11] A comparable desperation lies just below the surface of this remark by Carlos Baker: 'it would be quite without form were it not that Shelley has characteristically gathered his images under the roof of a ruling idea: the harmonizing power of love', Baker, 208.

Wittily Shelley points a contrast between the plodding slowness of Wordsworth's compositional rate and the nonchalant speed of his own. But the self-praise is neither bumptious nor ungracious. Shelley concedes Wordsworth's achievement in the *Lyrical Ballads*; however disrespectful, his association of Peter Bell with King Lear and Othello brings into play awareness of a kind of literature quite different from anything attempted by *The Witch of Atlas*. Of course, the link between *Lear* and *Peter Bell* which Shelley is making is that Wordsworth's handling of metre has a 'looped and windowed raggedness'. Shelley's conjunction of the two works is the most mocking of mock-compliments. Yet it suggests a blend of self-definition and self-questioning about the scope of *The Witch of Atlas*, a poem with only the most tangential relationship to the tragic. It is fair to assume that Shelley rejected the drift of Wordsworth's poem, its emphasis on the conversion of an unregenerate heart, as moralizing and dull. His answer in *The Witch of Atlas* is not to substitute a different, more acceptable moral, but to rely daringly on a self-evidently superior imaginative display. As Brian Nellist puts it: 'The paradoxical claim of the poem is that its playfulness makes it not less serious than *Peter Bell* and, indeed, more truthful.'[12] Shelley's is a 'visionary' trumping of Wordsworth's loyalty to 'the world of all of us, *and where | We find our happiness, or not at all*'.[13]

Wordsworth defines the central theme of *Peter Bell* in these lines:

> And now is Peter taught to feel
> That man's heart is a holy thing;
> And Nature, through a world of death,
> Breathes into him a second breath,
> More searching than the breath of spring.
>
> (1071–5)

This tribute to Nature's solacing power strikes one of the most resonant notes in the poem; the phrase which takes a characteristic Wordsworthian emphasis is 'man's heart'. Without denying Wordsworth's awareness of a 'world of death', Shelley creates in *The Witch of Atlas* his own self-validating, fictional world. For Wordsworth's

---

[12] 'Shelley's Narratives and "The Witch of Atlas"', 179.

[13] Wordsworth's phrases quoted—or, rather, slightly misquoted—with some scorn by Shelley in the Dedication to *Peter Bell the Third*, *PP* 324 (and in a late letter to John Gisborne, *Letters*, ii. 406).

'holy thing', he substitutes his half-mocking 'holy song' (538), though the holiness of *The Witch of Atlas* has nothing to do with traditional moral definitions. The Witch's voyage in the boat acts as an inspired retort to Wordsworth's humorously conscience-ridden timidity in the Prologue to *Peter Bell*.[14] There, Wordsworth turns away from the incitements of his aerial canoe, dismissing them as irrelevant to the poet who does not wish to 'quite forget | What on the earth is doing' (*Peter Bell*, 119–20). Yet Shelley's pursuit of 'pleasures'— 'Without impediment or let' (*Peter Bell*, 117 and 118)—travels beyond mere indulgence, has its own imaginative rigour. The intricate, demanding idiom of *The Witch of Atlas*—as Richard Cronin says, 'The world of the poem is the "noon of interlunar night" '[15]—reverses Wordsworth's kindly contempt for 'mystery':

> There was a time when all mankind
> Did listen with a faith sincere
> To tuneful tongues in mystery versed;
> *Then* Poets fearlessly rehearsed
> The wonders of a wild career.
>
> (*Peter Bell*, 121–5)

With its assumption that such 'faith sincere' is outmoded, this stanza prompts the visionary reply of Shelley's poem. The relationship between the two poems is a fascinating example of Shelley's critical engagement with the older poet's work. Ironically, Wordsworth's defence in the Prologue to *Peter Bell* of his kind of poetry anticipates the tactics used by Shelley; each poet affects modesty about his chosen course but insinuates claims for its importance. Wordsworth speaks of his poetry as proceeding along a 'lowly way' (138). In the previous stanza, however, he has put forward an eloquent apologia: 'The common growth of mother-earth | Suffices me—her tears, her mirth, | Her humblest mirth and tears' (133–5). There, the last line dwells, and makes the reader dwell, on the enduring contraries of existence. Shelley's way of making us aware of 'The common growth of mother-earth' is to transport us to a realm from which our usual concerns are made to appear more truly and more strange.

'So full of shapes is fancy, | That it alone is high fantastical':[16] Orsino's lines may smack of self-indulgence; they anticipate the

---

[14] See *SM* 174.   [15] 'Shelley's Witch of Atlas', 96.
[16] *Twelfth Night*, I. i. 14–15.

uncompromising detachment from human emotion shown by the
Witch. The charge of self-indulgent detachment is one which Shelley
ought to be dealing with in the introductory stanzas; they gently tease
Mary Shelley about her strictures on the poem, 'Upon the Score of
its Containing No Human Interest'. But Shelley sidesteps the issue;
he speaks as if Mary's objections to his verses were based on their
refusal to shape a didactic message from their fictions, on the fact
that 'they tell no story, false or true' (4). At once turning the tables
and sincerely complimenting, he deftly identifies Mary with his
Witch: an Ideal Being associated with Day's 'eternal smile':

> What hand would crush the silken-winged fly,
>   The youngest of inconstant April's minions,
> Because it cannot climb the purest sky
>   Where the swan sings, amid the sun's dominions?
> Not thine. Thou knowest tis its doom to die
>   When Day shall hide within her twilight pinions
> The lucent eyes, and the eternal smile,
> Serene as thine, which lent it life awhile.
>
>   (9–16)

The effect of these lines is complicated. There is grace as well as
irony in the suggestion that Mary's indifference to the claims of
Shelley's poem mirrors the Witch's aloofness from human feeling.
Day's smile may be 'eternal', but only lends 'life awhile' to the 'silken-
winged fly'. If Mary chooses, so the lines imply, this most vulnerable
of poems will wilt through lack of understanding sympathy on the
part of its most valued reader. At the same time, a counter-current
of good-natured mockery and surprisingly emphatic assertion
emerges: the allusion to Spenser's *Muiopotmos*[17] allies the poem to
a tradition of poetry which asserts 'the fragility and value of "visionary
rhyme"'.[18] Like *The Witch of Atlas* Spenser's poem, for all its grace
and sophistication, is deeply concerned with the vulnerability of
'innocence'. I surround the word with quotation marks because it
carries a slightly different charge for both poems. *Muiopotmos* is
sophisticated in its style and tone; its invoking of the 'Tragick Muse'
(413) is mock-epic rather than wholly serious. However, it does
lament the destruction of Clarion, the 'fresh yong flie' (33), by
Aragnoll in terms which are both moral and philosophical:

---

[17] See *SM* 169–71.   [18] *SM* 169.

But what on earth can long abide in state?
Or who can him assure of happie day;
Sith morning faire may bring fowle euening late,
And least mishap the most blisse alter may?

(217–20)

Shelley's sense of his poem's 'innocence' has to do more with its
imaginative freedom from constraints.

And yet, aware as he is of his poem's proneness to being mis-
understood, Shelley's manner in the introductory stanzas displays
a light but buoyant confidence. Refusing to theorize, his defence tries
to persuade by means of tone. The sense of hurt implicit in this
attempt slides into self-pity at the end of the third stanza: 'O, let
me not believe | That anything of mine is fit to live!' (23–4). Yet
the sudden cry works; it exposes the loneliness of this poem, with
nothing to recommend it except its sheer imaginativeness, its beauty.
Shelley has come a long and increasingly solitary way since *The
Revolt of Islam*, the far more polemical 'winged Vision' alluded to
in line 17 ('To thy fair feet a winged Vision came'), also preceded
by introductory stanzas to Mary: stanzas, though, which represent
the poet and his wife as comrades in ideological arms ('Two tranquil
stars' (124)).

Where *The Sensitive-Plant* admits a deep need to understand 'this
life | Of error, ignorance and strife' (Conclusion, 9–10), *The Witch
of Atlas*, no less conscious of that life, challenges its assumption of
primacy. The antithesis needs qualifying: the Conclusion of *The
Sensitive-Plant* makes a bid to alter the way we perceive 'this life';
The Witch of Atlas acknowledges what Harold Bloom calls the
'intolerable realities that art cannot mitigate'.[19] Still, *The Witch of
Atlas* is more a celebration of the imagination's power than a lament
over its limits. Unlike Peter in *Peter Bell*, the Witch undergoes only
the most hypothetical of conversions:

Tis said in after times her spirit free
   Knew what love was, and felt itself alone—

(585–6)

In these lines, the Witch seems to ascend and fall in the same stride
into another state of consciousness, one where she would 'know'

---

[19] 'The Unpastured Sea: An Introduction to Shelley', *The Ringers in the Tower*,
107.

what love was as a psychological need rather than 'be' love. Elsewhere in the poem, it would be more accurate to see the Witch as incarnating love in the sense of embodying the creative impulse, sometimes less the surrogate of the human poet than of that power which in various works Shelley describes as visiting the creative imagination.[20] Love as the need for sympathy is well understood by Shelley. In his essay *On Love* he sees love as prompted into being 'when we find within our own thoughts the chasm of an insufficient void and seek to awaken in all things that are, a community with what we experience within ourselves' (*PP* 473). But the desire for the self's ideal 'antitype' (*PP* 474), a desire arising from the hunger to escape loneliness, belongs to shadowy 'after times' only glanced at in lines 585–6. During the poem the Witch is 'alone' in a different sense: her 'spirit' is 'free' from human need; her 'aloneness' is the aloofness of a being unable to feel as mortal creatures feel. If we have sympathy for the Witch, it is the peculiar sympathy induced by the thought of her immunity to human suffering. 'Aloneness' in this second sense is presented by Shelley with wry humour, as in these earlier lines which describe Pan visiting the Witch: '[He] felt that wondrous lady all alone— | And she felt him upon her emerald throne' (119–20). Here the Witch is in her mythopoeic element, even as Shelley mingles a hint of plangency in the use of 'alone' with the lightest suggestion of an ethereal *double entendre* in the use of 'felt'.

Shelley's dealings with myth in *The Witch of Atlas* share in the poem's overall tonal sophistication. It is his lively, unsettling switches from the amusing to the serious which have to be borne in mind in weighing the worth of this description of the Witch: 'She is both myth herself, and the sophisticated modern poet's meditation upon myth, and, perhaps beyond that again, a projected *alter ego* to convey his anxieties about his own creativity.'[21] Central to the 'modern poet's meditation upon myth' in *The Witch of Atlas* is a playful sense of humour. As we have seen, Wordsworth in *Peter Bell* regards— with a certain regret—the old mythologies as exploded: '*Then* Poets fearlessly rehearsed | The wonders of a wild career'. In his opening stanza Shelley may seem to accept Wordsworth's mood of regret:

---

[20] See Cameron, 274–5.
[21] Marilyn Butler, *Romantics, Rebels and Reactionaries: English Literature and its Background 1760–1830* (Oxford, New York, Toronto, and Melbourne, 1981), 132.

Before those cruel Twins, whom at one birth
  Incestuous Change bore to her father Time,
Error and Truth, had hunted from the earth
  All those bright natures which adorned its prime,
And left us nothing to believe in, worth
  The pains of putting into learned rhyme,
A lady-witch there lived on Atlas' mountain
Within a cavern, by a secret fountain.

(49–56)

The poem begins by admitting its distance from the 'bright natures'
with which it seeks to re-establish imaginative links. But it admits
this in a distinctly mock-rueful way; by the end of the stanza we
have moved from the jadedly rational present in which obsession
with 'Error' and 'Truth' has 'left us nothing to believe in' to a fictional
past and the poet's own mythic creation. The stanza is full of witty
shiftings of tone. Shelley speaks of the 'pains of putting into learned
rhyme', yet the fluency of his *ottava rima* denies the 'pains' of
composition. The idea of the poem as something of a comic romp
is further sustained by the jokey description of the poem's central
fiction as a 'lady-witch'. However, a more imaginatively suggestive
note insinuates itself in the final line, where the 'cavern' and 'secret
fountain' invite, without attempting to satisfy, the reader's curiosity
about possible symbolic significance.

  Shelley's easy freedom at the outset of *The Witch of Atlas* may
appear to declare the poem's unimportance and thus to prove
the essential rightness of Wordsworth's rejection of myth. But the
following stanzas redress any such impression. They display a more
demanding inventiveness. What Shelley does is to demonstrate, in
Richard Cronin's words, 'the art of rising, not sinking, in poetry'.[22]
Without sacrificing their tonal flexibility, stanzas II to V offer
their own highly individual version, a version at once funny and
dignified, of Spenserian vision. The stanzas, as has often been
noticed, rework Spenser's account of the birth of Belphœbe and
Amoret after Chrysogonee has been 'wondrously' made pregnant
'Through influence of th'heauens fruitfull ray' (*The Faerie Queene*,
III. vi. 6):[23]

[22] 'Shelley's Witch of Atlas', 92.
[23] See *SM* 177–80.

Her mother was one of the Atlantides—
  The all-beholding Sun had ne'er beholden
In his wide voyage o'er continents and seas
  So fair a creature, as she lay enfolden
In the warm shadow of her loveliness . . .
  He kissed her with his beams, and made all golden
The chamber of grey rock in which she lay—
She, in that dream of joy, dissolved away.

(57–64)

The echo is not a symptom of a flagging in poetic inspiration before the poem has got properly airborne. Spenser's passage serves as a model and an incitement to self-definition for Shelley's poem, despite the element of parody in *The Witch of Atlas*. In describing the sexual encounter that leads to the Witch's birth, Shelley's diction—'the warm shadow of her loveliness', 'dream of joy'—idealizes its original. Spenser is more matter-of-factly intent on his story:

> The sunne-beames bright vpon her body playd,
> Being through former bathing mollifide,
> And pierst into her wombe, where they embayd
> With so sweet sence and secret power vnspide,
> That in her pregnant flesh they shortly fructifide.

(*The Faerie Queene*, III. vi. 7)

Yet Shelley does not coyly prettify; his language serves to direct us to the imagination which is presiding over the poem's transformations, its 'dissolvings away'. Bloom insists that the 'allusion tells us that the Witch has been born into the Garden of Adonis'.[24] But it is less the thematic relevance of the allusion which matters than what it reveals about Shelley's involvement with the business of fictionalizing. Spenser loses himself in the strange beauty of the episode; Shelley makes us aware that his lines are a nimble-witted piece of story-telling, that the story-teller is both amused and fascinated by the process of inventing myth.

In place of the 'pleasures' that Wordsworth would deny himself, the third stanza offers a more extreme and comic virtuosity:

> 'Tis said, she first was changed into a vapour,
>   And then into a cloud, such clouds as flit,

[24] *SM* 180.

Like splendour-winged moths about a taper,
  Round the red West when the sun dies in it:
And then into a meteor, such as caper
  On hill-tops when the moon is in a fit:
Then, into one of those mysterious stars
Which hide themselves between the Earth and Mars.

(65–72)

Again, the tones are complex, a consequence of Shelley's mobile use of syntax. It is true that 'The dignity of the Witch is constantly kept at the mercy of Shelley's language',[25] but Shelley's language is deeply engaged by the material which it parodies. Assuredly, the spate of similes makes fun of the symbolic idiom which the poet of *Prometheus Unbound* found necessary. However, there is a great deal of vitality, too. Shelley is continuing, albeit with more wit than usual, to write a poetry that brings its own reality into being. Ridicule neither unbalances the affirmative energies of the poem, nor is it merely a means to explore sceptically the function of poetry. The poet's teasing of his imaginative creation often sharpens as he delights in the licence which she allows him. At times, Shelley anticipates the more aggressively comic vision of stanzas XXVII and XXVIII, say, from Byron's *The Vision of Judgement*:

As things were in this posture, the gate flew
  Asunder, and the flashing of its hinges
Flung over space an universal hue
  Of many-colour'd flame, until its tinges
Reach'd even our speck of earth, and made a new
  Aurora borealis spread its fringes
O'er the North Pole; the same seen, when ice-bound,
By Captain Parry's crew, in 'Melville's Sound.'

And from the gate thrown open issued beaming
  A beautiful and mighty Thing of Light,
Radiant with glory, like a banner streaming
  Victorious from some world-o'erthrowing fight:
My poor comparisons must needs be teeming
  With earthly likenesses, for the night
Of clay obscures our best conceptions, saving
Johanna Southcote, or Bob Southey raving.

[25] 'Shelley's Witch of Atlas', 91.

As Jerome J. McGann remarks: ' "The Witch of Atlas" and "The Vision of Judgement" seize a visionary tradition far removed from the visionary emphasis of *The Prelude* and "Kubla Khan," and variant from "Adonais" and *Childe Harold* III as well.'[26] One has to add to McGann's assertion the fact that *The Witch of Atlas* and *The Vision of Judgement* are very different from one another. Though the way they are written shows a common dislike for what Shelley calls the 'over busy gardener's blundering toil' (32), there is a sharp contrast between the two poems in their attitudes to the 'visionary' and their uses of comedy. At bottom, Shelley's inventive-ness challenges, but Byron's reinforces, common sense; in *The Witch of Atlas* the comic serves to beguile the reader into accepting the seemingly skew-whiff perspectives of 'visionary rhyme'; in *The Vision of Judgement* the comic serves a more sardonic, satirical, *ad hominem* function.

That said, both poems display a comparable freedom and energy. In both the poet as inspired seer gives way to the poet as virtuoso. In both far more of the ordinary self persists into 'vision' (however humorously the visionary is presented). So, in the stanzas just quoted from *The Vision of Judgement*, Byron frivolously sends up, but participates in, a visionary rhetoric. His first stanza seems intent on subversion. Yet the final couplet is amusingly unexpected. Half seriously, half mock-seriously, Byron substantiates the reality of the 'universal hue | Of many-colour'd flame' by switching the focus of his attention from celestial phenomena to 'Captain Parry's crew'. The rhythm, expecting a clash between vision and fact, is calmed by their surprising truce. Again, in the second stanza, Byron admits the splendour of 'A beautiful and mighty Thing of Light', even as 'Thing' awaits comic exploitation. When it comes, in what could be a parody of Shelleyan similes, the writing still manages to relate the insult of the last line to a more than debunking glance at the visionary poet's problems. There is a touch of sympathy in '*our* best conceptions' (my italics). In other words, neither Byron nor Shelley, for all the differences between their poems, is limited by the comic mode. In the famous exchange between Satan and Michael in stanza XXXV of *The Vision of Judgement*, for example, the comic expands to include a moment of drama that is both droll and poignant:

[26] 'A Reply to George Ridenour', one of a number of short articles, 'On Byron', *SIR* xvi (1977), 576.

though they did not kiss,
Yet still between his Darkness and his Brightness
There pass'd a mutual glance of great politeness.

Timothy Webb helpfully distinguishes, then, between the two poets: 'Byron used the comic largely as a weapon, Shelley used it largely as a defence.'[27] Shelley is defending, among other things, against the charge of visionary pretentiousness. Yet his comic defence, his pose of detachment from his creation, gives rise to a poetry which is often more realized than the lyricism of *Prometheus Unbound*. The Fourth Spirit's Song in the lyrical drama shares the same theme as this stanza, the sixteenth, from *The Witch of Atlas*:

And odours in a kind of aviary
  Of ever blooming Eden-trees she kept,
Clipt in a floating net a love-sick Fairy
  Had woven from dew beams while the moon yet slept—
As bats at the wired window of a dairy
  They beat their vans; and each was an adept,
When loosed and missioned, making wings of winds,
To stir sweet thoughts or sad, in destined minds.

(169–76)

Though both passages attempt to express the idea of 'Nurslings of immortality' (*Prometheus Unbound*, I. 749), their tones diverge. The poet of *Prometheus Unbound* refrains from a physical evocation of 'Forms more real than living man' (I. 748). The poet of *The Witch of Atlas* sketches a recklessly parodic inner world. Arguably, stanza XVI lacks the intellectual discipline of the lyrical drama. But it has great zest, as it gives life to the incorporeal by bringing it into contact with the ordinary: those 'bats at the wired window of a dairy', for example. For all the vigour of the writing, however, Shelley does not settle for the merely concrete or physical. There is an air of high spirits, a delight in yoking together the incongruous, which mirrors the entire poem's concern with juxtaposition of opposites: humour and seriousness, immersion in self-created myth and laying bare the process of fictionalizing, visionary rhyme and human interest. Shelley creates a poem with its own predominant texture and mood, yet he does not smooth over gaps between tones, styles, perspectives. To the contrary, he makes a point of drawing the reader's attention to

---

[27] *The Violet in the Crucible: Shelley and Translation* (Oxford, 1976), 130.

such gaps, doing so in a way which manages to seem uncannily self-aware rather than irritatingly clever-clever. So in stanza XIV he confronts the reader with dizzying verbal high jinks and an abrupt, touching change of mood:

> The deep recesses of her odorous dwelling
>   Were stored with magic treasures—Sounds of air,
> Which had the power all spirits of compelling,
>   Folded in cells of chrystal silence there;
> Such as we hear in youth, and think the feeling
>   Will never die—yet ere we are aware,
> The feeling and the sound are fled and gone,
> And the regret they leave remains alone.
>
>   (153–60)

The stanza begins with an intricate interplay between sensuous and abstract imagining: in the description of 'Sounds of air' being 'Folded in cells of chrystal silence', the language at once glimpses and yields up the possibility of capturing 'chrystal silence'. Such writing emphasizes the poem's otherness, its status as a performance which gives delight, but stays remote from the reader's concerns. It is this remoteness which the ensuing lines, with their switch to a less figurative, more emotional style, challenge. The 'Sounds of air' come to represent feelings experienced in youth, lost in later years; the writing would risk seeming sentimental were it not for the creative jostling between idioms and viewpoints.

Shelley's use of mythology reveals a comparable escape from the graceful confines which it appears to set for itself. Stanzas VI to XI which catalogue figures from the old mythologies toy with pastiche, yet go a long way beyond it. In stanza VIII, for instance, Shelley does two things: he covertly re-establishes the reader's interest in myth and he insinuates the superiority of his own mythic creation:

> And old Silenus, shaking a green stick
>   Of lilies, and the wood-gods in a crew
> Came, blithe, as in the olive copses thick
>   Cicadæ are, drunk with the noonday dew:
> And Dryope and Faunus followed quick,
>   Teasing the God to sing them something new
> Till in this cave they found the lady lone,
> Sitting upon a seat of emerald stone.
>
>   (105–12)

Stanza after stanza finishes as this one does, modulating from sly humour into wonder. The effect is of an effortless crescendo culminating in the affirmation of stanza XII:

> For she was beautiful—her beauty made
>   The bright world dim, and every thing beside
> Seemed like the fleeting image of a shade:
>   No thought of living spirit could abide—
> Which to her looks had ever been betrayed,
>   On any object in the world so wide,
> On any hope within the circling skies,
> But on her form, and in her inmost eyes.

                                                    (137–44)

Here, Shelley pays homage to the source of his fictions. His language is content to withdraw into a haven of abstraction from which 'every thing beside | Seemed like the fleeting image of a shade'. The variegated world of mythology resigns itself to the superior beauty of the poem's own myth. This stanza broaches that quarrel or friction between the imaged and its imageless source which makes of *The Witch of Atlas* so much more than a mere exercise of the fancy. These lines authenticate their assertions—'For she was beautiful—her beauty made | The bright world dim'—partly through their rhythmic sureness, partly through the fact that the rest of the poem is so alive to the beauty of the 'bright world'. 'Betrayed' carries a faint but unmistakable overtone of menace, foreshadowing the fate of Rousseau after his confrontation with the ambiguous 'shape all light' in *The Triumph of Life*. As the next stanza reveals, however, with its account of the Witch's weaving of a veil to protect onlookers from her unendurable beauty, this subliminal menace does not turn into a presiding presence.

Indeed, the menace of the Witch's beauty is exorcized—or, at least, qualified—in the stanzas (XXII–XXIV) in which she confronts the mythological creatures who wish to 'live forever in the light | Of her sweet presence' (223–4):

> The Ocean-Nymphs and Hamadryades,
>   Oreads and Naiads with long weedy locks,
> Offered to do her bidding through the seas,
>   Under the earth, and in the hollow rocks,
> And far beneath the matted roots of trees
>   And in the knarled heart of stubborn oaks,

So they might live forever in the light
Of her sweet presence—each a satellite.

'This may not be—' the wizard Maid replied;
  'The fountains where the Naiades bedew
Their shining hair at length are drained and dried;
  The solid oaks forget their strength, and strew
Their latest leaf upon the mountains wide;
  The boundless Ocean like a drop of dew
Will be consumed—the stubborn centre must
Be scattered like a cloud of summer dust—

'And ye with them will perish one by one—
  If I must sigh to think that this shall be—
If I must weep when the surviving Sun
  Shall smile on your decay—Oh, ask not me
To love you till your little race is run;
  I cannot die as ye must . . . over me
Your leaves shall glance—the streams in which ye dwell
Shall be my paths henceforth, and so, farewell!'

                                          (217-40)

Shelley wins pathos and comedy from his account of the Witch's
inability to share the limitations of mortality. The pathos is not
merely the consequence of the gulf between human and non-human
perspectives, as it is in 'The Cloud' where 'Through the murk of
our epistemological twilight the spirit of the cloud may appear
to taunt us'.[28] Here, it is the non-human, surrogate deity's awareness of contradictions, caught in stanza XXIV's 'modulations of
her struggle with herself',[29] which is affecting. In these stanzas
the ambiguities surrounding the Witch come to a sharp focus.
What moves her is the thought of her invulnerability to the fate the
nymphs will eventually suffer. 'I cannot die as ye must' is ironic in
that it represents a strength as though it were (which, in a sense,
it is) a flaw. On the other hand, and this is where comedy asserts
itself, there is no mistaking the Witch's settled decisiveness. A
comic undertone offsets the sobbing grief of 'and so, farewell'. The
gesture—the words both are, and suggest, a gesture—is closer to
the histrionic than the tragic. As so often in *The Witch of Atlas*,
Shelley unsettles the reader by allowing tones and styles to collide:

---

[28] Timothy Webb, *Shelley: A Voice Not Understood* (Manchester, 1977), 250.
[29] *SM* 190.

grief jostles with humour, pathos rubs shoulders with self-regard, sympathy clashes with detachment. Stanza XXIII's comic apocalypse exhibits Shelley's mastery over his materials: 'the stubborn centre must | Be scattered like a cloud of summer dust'. The mastery is such that the writing threatens to leave the reader unmoved. The poetic impact is richly indeterminate; we hear the sombre reverberations, and yet they are muffled by the very patness of rhyme and rhythm. There is a quick erasure of distress in the following lines, stanza XXV:

> She spoke and wept—the dark and azure well
>   Sparkled beneath the shower of her bright tears,
> And every little circlet where they fell
>   Flung to the cavern-roof inconstant spheres
> And intertangled lines of light—a knell
>   Of sobbing voices came upon her ears
> From those departing Forms, o'er the serene
> Of the white streams and of the forest green.
>
> (241-8)

As the language relishes the imaginative possibilities of the Witch's tears, their 'inconstant spheres | And intertangled lines of light'— lines where geometry gets the better of sorrow—so the 'departing Forms' are gently ushered out of the poem. 'Forms' is, as Bloom claims, 'wonderfully apt'.[30] In saying this, Bloom probably has in mind Spenser's description of the Garden of Adonis (a passage central to his reading of The Witch of Atlas) where 'The substance is not chaunged', but 'formes are variable and decay'. He is sensitive to Shelley's individual, non-Platonic use of a word commonly associated with Plato's realm of ideal essences: 'In the Garden of Adonis, as in Blake's Beulah or the state of existence embodied in "The Witch of Atlas," forms constantly pass away, but substance always endures to be converted into new forms again.'[31] Here Bloom presses the word more strenuously for metaphysical implications than I would wish to do. My reading is more interested in the skill with which Shelley blends sadness with aloof detachment—'Forms' just glances at the poem's central, never explicitly stated, concern with imaginative inventions. Departing, these mythological figures grow ghostly, insubstantial: superseded fictions. It is in stanza XXVII, however,

[30] SM 191.   [31] SM 179.

that Shelley finds an image that does justice to his poem's movement
into the chilling splendours of vision:

> While on her hearth lay blazing many a piece
>   Of sandal wood, rare gums and cinnamon;
> Men scarcely know how beautiful fire is—
>   Each flame of it is as a precious stone
> Dissolved in ever moving light, and this
>   Belongs to each and all who gaze upon.
> The Witch beheld it not, for in her hand
> She held a woof that dimmed the burning brand.
>
> (257–64)

The image is of fire; it anticipates the condition Wallace Stevens
imagines as a 'diamond jubilance beyond the fire' in the third section
of 'The Owl in the Sarcophagus'. Shelley presents this fire with all
the resources of an art that clings to the image, but acknowledges
its final ineffectuality. So visual yet so visionary, 'a precious stone
| Dissolved in ever moving light' is at once beyond the reach of the
reader's senses and the possession of 'each and all who gaze upon'.
And, crucially, the Witch's 'pictured poesy' which she was 'broidering'
(252) in the previous stanza now provides the 'woof that dimmed
the burning brand'. Again, the poetry conveys its suggestions
too richly for abstract translation to be anything but reductive;
nevertheless, the reader approaches here the heart of the poem: the
exhilarating duel it stages between 'inspiration' and 'composition',
between 'vision' and its imaged manifestation in 'poetry'.

Where in *Alastor* or *The Triumph of Life* Shelley's language enacts
the elusiveness of his perceptions and in *Prometheus Unbound* relies
for its force on cumulative effects of patternings and echo, it delights
in *The Witch of Atlas* in a local brio. Even his use of reflexive imagery,
the most obvious token of his concern with the imaginative process,
is different in this poem. In *Alastor* the way such images work often
mimics the Poet's pursuit of an ideal self-image. So in these lines,
'Her voice was like the voice of his own soul | Heard in the calm
of thought' (153–4), the wording threatens to collapse the distinction
between 'voices' which 'like' appears to herald. The reader is offered
little to apprehend sensuously in the lines, is conducted, rather, into
a labyrinth of subjectivity from which there is no escape. The
following passage from *The Witch of Atlas*, stanza XXXIII, also
employs a reflexive image, but the contrast with *Alastor* is strong:

The plant grew strong and green—the snowy flower
  Fell, and the long and gourd-like fruit began
To turn the light and dew by inward power
  To its own substance; woven tracery ran
Of light firm texture, ribbed and branching, o'er
  The solid rind, like a leaf's veined fan—
Of which Love scooped this boat—and with soft motion
Piloted it round the circumfluous Ocean.

(305–12)

Though Shelley's physical precision laughs at the non-physical essence of his subject—its 'inward power'—it illustrates deftly how that power manifests itself. This is partly the result of the fluently handled syntax which mirrors the 'ribbed and branching' process of growth. The stanza is both a vindication and sophisticated mockery of Romantic organicism. Shelley might seem closer in such a stanza to 'fancy' than 'imagination'. But one of the strengths of *The Witch of Atlas* is the paradoxical freedom to imagine which 'fancy' gives the poet. If stanza XXXII with its clumsy humour—'The first-born Love' (298) compared to a 'horticultural adept' (300)—reveals the lax use to which this freedom is at times put, stanza XXXIII typifies the energy it more frequently generates. Poised between intuiting the plant's 'substance' as 'inward power' and as outward 'texture', the writing struggles profitably between its assured yet teasing physical reality and its approach to allegory. 'Love'—an abstraction which elsewhere functions as the principle behind the poem's fictions—changes character in the two stanzas, XXXII and XXXIII. In the first, Love is a coyly mythological Cupid; in the second, an unexpectedly forceful surrogate—'Love scooped this boat'—of the creative mind. Usually reflexive imagery redirects attention to the tenor of a comparison, yet in *The Witch of Atlas* it is as interested in the vehicle. Throughout the poem, Shelley respects, does not grow impatient with, the imaginative life of his fictions.

This respect encourages a new kind of visionary momentum in Shelley's language. The boat journey on which Hermaphroditus and the Witch embark is so much both a creation in its own right and a way of describing the creative process that the reader may be particularly tempted to describe this section of the poem as 'reflexive'. Indeed, William Keach has called the poem 'an extended, self-delighting display of the imagination's ability to project itself as other and to invest a fictional natural setting with its own reflexive operations'.[32] In *Alastor* the Poet's journey conducts him through a landscape which struggles to move from the picturesque into the symbolic. At certain moments, such as the contrast which presents itself between the downward-looking flowers and the inward-looking poet, this struggle is resolved, and landscape detail achieves a resonance that has nothing to do with mere embellishment. The landscape traversed by the Witch is neither conventionally picturesque nor earnestly symbolic. In a stanza such as XLVI, it is a mocking but authentic example of the visionary—heir to Coleridge's 'Kubla Khan', forebear of Yeats's Byzantium poems:

> The water flashed, like sunlight by the prow
>   Of a noon-wandering meteor flung to Heaven;
> The still air seemed as if its waves did flow
>   In tempest down the mountains—loosely driven
> The lady's radiant hair streamed to and fro:
>   Beneath, the billows having vainly striven
> Indignant and impetuous, roared to feel
> The swift and steady motion of the keel.

                                        (409–16)

Here, similes offer exact visionary equivalents for the movement of the boat, justifying their presence. In its transition from wayward energy to ultimate purpose, the second line imitates the behaviour of Shelley's descriptive poetry at its greatest. The next two lines enact another transition—the air is 'still' but 'seemed as if its waves did flow'—and point up the poem's delight in figurative transformations. This delight assumes breathtaking force in the final thrust of 'visionary cynicism'.[33] Bloom has suggested a comparison between Shelley's 'indignant' billows and Yeats's 'uncontrollable sea'.[34] Yet the emphases of the two poets diverge. For all the 'self-awareness of limitations'[35] displayed by *The Witch of Atlas*, its greatness lies in its refusal to admit obstacles; the vain struggle of the billows against the progress of the Witch's boat suggests the superiority of the visionary imagination to the natural world; the impact of the lines is exhilarating from one perspective, the Witch's, sobering from another, the human. But they do not reveal that passionate involvement with the 'gong-tormented sea' of turbulent passion which

[32] Keach, 111.   [33] *SM* 165.   [34] *SM* 201.   [35] *SM* 198.

Yeats's 'Byzantium' ends up exhibiting, almost despite itself and its imagining of escape from 'unpurged images'.

Ultimately, *The Witch of Atlas* shows a desire to break through any 'self-awareness of limitations'. An intuition of this desire may have prompted Hazlitt's preference for the journey in *The Witch of Atlas* over that in *Alastor*: 'We cannot help preferring *The Witch of Atlas* to *Alastor, or the Spirit of Solitude*; for, though the purport of each is equally perplexing and undefined, (both being a sort of mental voyage through the unexplored regions of space and time), the execution of the one is much less dreary and lamentable than that of the other.'[36] While Hazlitt goes on to temper this already qualified praise—'In the "Witch," he has indulged his fancy more than his melancholy, and wantoned in the felicity of embryo and crude conceits even to excess'[37]—he pinpoints the exuberance with which Shelley voyages through the 'unexplored regions of space and time', an exuberance which wears a nonchalant mask and reaches a crescendo in the description of the Witch's assault on death. Shelley starts with an echo of the 'Paradise of vaulted bowers' (*Prometheus Unbound*, II. v. 104) into which Asia's lyric, 'My soul is an enchanted Boat' leads the reader: 'and the grave | Of such, when death oppressed the weary soul, | Was as a green and overarching bower | Lit by the gems of many a starry flower' (597–600). But he counterpoints that dream with an undreamlike display of power:

> For on the night that they were buried, she
>   Restored the embalmer's ruining, and shook
> The light out of the funeral lamps, to be
>   A mimic day within that deathly nook;
> And she unwound the woven imagery
>   Of second childhood's swaddling bands and took
> The coffin, its last cradle, from its niche
> And threw it with contempt into a ditch.
>
> And there the body lay, age after age,
>   Mute, breathing, beating, warm and undecaying
> Like one asleep in a green hermitage
>   With gentle smiles about its eyelids playing
> And living in its dreams beyond the rage
>   Of death or life, while they were still arraying

[36] From a review of Shelley's *Posthumous Poems*, in Redpath, *The Young Romantics and Critical Opinion*, 394.
[37] Redpath, *The Young Romantics and Critical Opinion*, 394.

In liveries ever new, the rapid, blind
And fleeting generations of mankind.

(601–16)

In these lines, Shelley both hints at and overrides the imagination's
limitations. The Witch is described as redeeming the false 'imagery'
of the real world. Yet the body she resuscitates through her artifices
inhabits a 'green hermitage' which has the ring of a limbo or
never-never land. A comparable ambivalence hovers round this
body and Hermaphroditus whose 'unawakened' state is described
in similar terms: 'And ever as she went, the Image lay | With
folded wings and unawakened eyes; | And o'er its gentle countenance
did play | The busy dreams' (361–4). The body at once represents
a triumph over death and a fantasized escape from the difficulties
of living, just as Hermaphroditus is both super- and non-human,
obviously crucial to the Witch's designs, yet too comically presented
for Kenneth Cameron's allegorizing gloss to convince: 'the relationship of the hermaphrodite and the Witch is similar to that of
the creative imagination and the power in *A Defence of Poetry*'.[38]
However regretfully or contemptuously, the final lines of stanza
LXXI restore the unignorable reality of 'the rapid, blind | And
fleeting generations of mankind', lines in which there is a switch
of energy from the Witch to her adversary. Shelley describes the
body as 'living in its dreams', a phrase which gathers to itself
the two-edged meaning of 'dream' in this poem: for those visited
by the Witch, 'dreams' are an avenue of escape from 'reality', yet
'dreams' seem a respite from the Witch's own kind of exhilarating
activity.

Indeed, as the poem continues, it modulates: from the realm
of vision, Shelley declines into graceful, witty wish-fulfilment. It is
in these stanzas (LXXII–LXXVII) that poetic invention reveals its
special alliance with scepticism. When Shelley delights in the exposure
of 'How the god Apis, really was a bull | And nothing more' (627–8),
but two stanzas later mockingly abets the somnambulistic transformation of soldiers into blacksmiths into Cyclopses (641–8),
one is struck by a paradox to which Hazlitt drew attention: 'Indeed
it is curious to remark every where the proneness to the marvellous
and supernatural, in one who so resolutely set his face against every

---

[38] Cameron, 274.

received mystery, and all traditional faith.'[39] It is easy in hindsight to identify this 'proneness to the marvellous and supernatural' with the promptings of the Romantic Imagination. In my view, however, it is more accurate to posit a connection between the multiplicity of the poem's fictions and the imaginative spur provided by a scepticism which Hazlitt, disliking Shelley's way of advertising his radicalism as willed and irresponsible, called 'obnoxious and indiscreet',[40] but which strikes me as generously open-minded. That the presence of complicating drives makes the poem hard to paraphrase cannot be denied. Much of *The Witch of Atlas* might seem to answer to Douglas Bush's description: 'He will, unlike our contemporaries, make his principal conception fairly clear, and then he feels at liberty to adorn or obscure it with a multitude of details which, we often feel sure, have significance for him but remain hazy to us.'[41] This describes what I regard as the main difficulty posed by *Prometheus Unbound*. But it also, despite itself, suggests the unique achievement of *The Witch of Atlas*, its capacity to realize imaginatively 'a multitude of details'. If anything remains 'hazy' about these details it is, as I have argued, their place in any imposed allegorical scheme. The poem's texture is superbly finished. Nor does Bush bring out what is the most impressive strength of *The Witch of Atlas*: its sophisticated control of tone.

The poem is best seen as a Shelleyan version of that activity of the spirit which Wordsworth describes in these lines from Book IV of *The Excursion*:

> As the ample moon,
> In the deep stillness of a summer even
> Rising behind a thick and lofty grove,
> Burns, like an unconsuming fire of light,
> In the green trees: and, kindling on all sides
> Their leafy umbrage, turns the dusky veil
> Into a substance glorious as her own,
> Yea, with her own incorporated, by power
> Capacious and serene.

(1062–70)

[39] Redpath, *The Young Romantics and Critical Opinion*, 394.
[40] Redpath, *The Young Romantics and Critical Opinion*, 394.
[41] *Mythology and the Romantic Tradition in English Poetry* (1937; New York, 1957), 139.

Yet Shelley's last stanza, with its promise of 'A tale more fit for the weird winter nights | Than for these garish summer days, when we | Scarcely believe much more than we can see' (670–2), refuses to abide in 'the deep stillness of a summer even'. There is a touch of wryness, even sadness in the final line's recognition of our enchainment to the evidence of our 'organs'. By means of fictions, metamorphic spells, *The Witch of Atlas* has tried to free its readers from that enchainment. Shelley's poem converts reality 'Into a substance glorious as her own'; with wit and tact, it counts the cost of doing so.

7

Epipsychidion
The Before Unapprehended
Relations of Things

It is sometimes felt that in *Epipsychidion* Shelley 'cheats' by writing a poem which celebrates the principle of loving as many people as possible, but which also selects a single person (Emilia Viviani) for special homage. This tension, however, is not an unrewarding or self-deceiving one. The poet's feelings about a relationship are twinned in complicated ways with his feelings about relationship itself. In the more polemical section of the poem (147–89), Shelley advocates the multiplying of affection; in its most compelling passages he strives to break down barriers between self and other, word and object. Such striving—rhapsodic yet subtle—fails but also accounts for the poem's success.

Throughout *Epipsychidion*, Shelley approaches what Foucault describes as a major Western enterprise: 'transforming sex into discourse'.[1] The consequences of this are complex. Shelley substitutes his own verbal universe for the 'real' world. *Epipsychidion*'s attitudes cannot be analysed in cold blood; they are always provisional, always in the process of being redefined. And yet the poem's autonomy is never absolute. *Epipsychidion* is enriched by the resistance its subject offers to its language. Admittedly, the poet's posture of adoring rapture goes hand in hand with a quite systematic attempt to erode the distinction between desire and reality until, as he phrases the idea in 'Fragments Connected with *Epipsychidion*', 'we | Become one being with the world we see' (185–6, *P W* 430). Such an erosion can be seen in these early lines in which Shelley tries to find an appropriate image for Emilia:

> Sweet Benediction in the eternal Curse!
> Veiled Glory of this lampless Universe!

---

[1] *The History of Sexuality, i: An Introduction*, trans. Robert Hurley (1978; Harmondsworth, 1984), 20.

> Thou Moon beyond the clouds! Thou living Form
> Among the Dead! Thou Star above the Storm!
> Thou Wonder, and thou Beauty, and thou Terror!
> Thou Harmony of Nature's art! Thou Mirror
> In whom, as in the splendour of the Sun,
> All shapes look glorious which thou gazest on!
>
> (25–32)

The writing here acknowledges its own rhetoric as it pulls, magician-like, image after image out of the hat. Yet it does not allow that self-consciousness to quench its intensity. Each image represents a new attempt and a new failure to define completely. The passage yields a paradoxically double sense that words are inadequate and that they manage adequately to tell us this. They convince us of the 'splendour' of their subject. The final image suggests a reciprocity between poem and woman. Shelley's wording—the fact that Emilia is both gazed upon and gazing—undercuts the suspicion of 'projecting' which 'Mirror' raises. The lines at once consider and subvert the charge of poetic narcissism.

Nevertheless, for all its attempts to blur distinctions, Shelley's language throughout *Epipsychidion*'s most impressive passages remains conscious of otherness: of the outer world, of separate though beloved selves. So when he writes. 'Ah me! | I am not thine: I am a part of *thee*' (51–2), the hopeful assertion expects to arouse disbelief. Yet the desire to do away with separateness by redefining the self and what lies outside it propels *Epipsychidion*. Shelley's graceful lines at the head of the poem—'almost', he writes in the *Advertisement* (*PP* 373), 'a literal translation from Dante's famous Canzone' (from the *Convito*, Trattato II)—bid farewell to an outmoded manner. When he writes, 'My Song, I fear that thou wilt find but few | Who fitly shall conceive thy reasoning, | Of such hard matter dost thou entertain', his tone is urbane, as though he were a subtle medieval allegorist and not, as his poem shows him to be, a poet who forfeits the discursive underpinning available to Dante. The poem proper is demanding in a different way: Shelley's words do not draw support from some external belief-system; instead, they seek to create their own values:

> Aye, even the dim words which obscure thee now
> Flash, lightning-like, with unaccustomed glow;

I pray thee that thou blot from this sad song
All of its much mortality and wrong,
With those clear drops, which start like sacred dew
From the twin lights thy sweet soul darkens through,
Weeping, till sorrow becomes ecstasy:
Then smile on it, so that it may not die.

(33–40)

Beneath the extravagant, courtly pose, there is a good deal of energy. It is apparent in the way Shelley vexes passion with passionately explicit scruples about language. It is there in line 34 which, packing its first two feet with stresses, vindicates its claim to possess an 'unaccustomed glow'; and it is manifest as the force which drives the poem through its attack on dualism and separateness. Shelley's paradoxes (those 'twin lights', for example, 'thy sweet soul *darkens* through') blend intensity and virtuosity in a way which deprives the words of their usual meaning. So, in the phrase—'till sorrow becomes ecstasy'—the reader is hard-pushed (yet the strain is felt as a creative one) to identify the feeling. Here, as elsewhere, *Epipsychidion* is more than a 'Romantic caricature of Petrarch', as Charles Tomlinson has called it.[2] Rather, it achieves the type of poetic life Shelley was to describe in *A Defence of Poetry*. Discussing 'poets', he would write: 'Their language is vitally metaphorical; that is, it marks the before unapprehended relations of things, and perpetuates their apprehension' (*PP* 482). Yet, thoroughgoing as Shelley's assault on the referential is, we are rarely tempted to dismiss *Epipsychidion* as merely verbal.

In his *Advertisement* at the head of the poem (*PP* 373), Shelley repeats Dante's scorn for those whose rhetorical 'colouring' does not screen 'a true meaning',[3] but the poem is not written in the translatable allegorical idiom which this use of Dante might seem to imply. Shelley's images are more conscious of their place in a would-be self-sustaining scheme, more conscious, too, of the difficulty of achieving that exact fit between word and object which constitutes 'true meaning'. Often the poetry mimes the fleeting nature of

---

[2] *The Times Literary Supplement*, 22 Apr. 1977, 475.
[3] Shelley quotes from *La Vita Nuova*, section XXV, the relevant passage from which is rendered as follows in *La Vita Nuova*, trans. Barbara Reynolds (1969; Harmondsworth, 1975): 'for it would be a disgrace if someone composing in rhyme introduced a figure of speech or rhetorical ornament, and then on being asked could not divest his words of such covering so as to reveal a true meaning', 75.

inspiration described in *A Defence of Poetry* ('the mind in creation is as a fading coal', *PP* 503–4). The pathos instinct in its greatest moments is well caught in a phrase used by D. J. Hughes when he speaks of 'the arrestation of the fading of the coal'.[4] *Epipsychidion* begins self-deprecatingly, protesting the sincerity of its 'votive wreaths' (4), but suggesting that an original glow of creativity has, indeed, faded:

> Sweet Spirit! Sister of that orphan one,
> Whose empire is the name thou weepest on,
> In my heart's temple I suspend to thee
> These votive wreaths of withered memory.

> (1–4)

However, it is the poem's achievement to transform 'withered memory' into living inspiration. At the end of one firework-display of figures (56–69), Shelley concludes: 'I measure | The world of fancies, seeking one like thee, | And find—alas! mine own infirmity' (69–71). Here 'measure' provides a steadying ballast. Apparently refuted by the poet's virtuosity, 'infirmity' flirts with a courtly idiom, knows it is doing so and asks to be read seriously, being both a confession of inadequacy and a sign of Shelley's determination to pursue his vision of a perfect match between word and object, self and other.

Is such a vision a mirage? The reader asks the question now in relation to Shelley's dealings with language, now in relation to the poet's feelings about a person. One of the passages which Shelley noted in Dante's *Convito*, 'Love, taken in the true sense, and subtly considered, is no other than the spiritual union of the soul and the thing beloved',[5] defines a philosophical position to which he is attracted as his title shows (Reiman suggests it means 'externalized little soul').[6] Yet it is a position which involves more clearly demarcated notions about the 'spiritual' than Shelley possessed. Moreover, if union with what (in the sense of the title) represents the best impulses of the self is desirable, it also poses a threat to the finite

---

[4] 'Coherence and Collapse in Shelley, with Particular Reference to *Epipsychidion*', *ELH* xxviii (1961), 267.

[5] *'The Banquet' (Il Convito) of Dante Alighieri*, trans. Katharine Hillard (London, 1889), 137. For Shelley's memoranda on Dante's *Convito* see Neville Rogers, *Shelley at Work: A Critical Inquiry* (2nd edn., Oxford, 1967), 340–1.

[6] *Percy Bysshe Shelley* (1969; London and Basingstoke, 1976), 125.

selves of the poet and his beloved. In the central autobiographical
sections (discussed below), Shelley makes fine poetry out of his
wavering view of Emilia's relationship to some Ideal. Yet, as 'Youth's
vision thus made perfect' (42), she is in danger of dissolving into
a shadowy, transcendental ideal or a dream-projection. In the long
passage, lines 190–216, the language suggests that the 'Being' whom
Shelley was haunted by in his youth was created by subjective need:
'In solitudes | Her voice came to me through the whispering woods'
(200–1). This Being is later described as projected outwards, then
quested for, as the poet asks 'Whither 'twas fled, this soul out of
my soul' (238). Emilia is the fulfilment of these intimations. But
Shelley swings throughout between seeing her as an embodiment of
an ideal and admitting that the significance she possesses is bestowed
upon her by his imagination. These uncertainties and qualifyings
improve the poem; they complicate and subtilize its vision of love.

The poem continually tries to narrow the scarcely acknowledged
gap between the poet and Emilia. The following questions, for
instance, screen and even dissolve her identity:

> Art thou not void of guile,
> A lovely soul formed to be blest and bless?
> A well of sealed and secret happiness,
> Whose waters like blithe light and music are,
> Vanquishing dissonance and gloom? A Star
> Which moves not in the moving Heavens, alone?
> A smile amid dark frowns? A gentle tone
> Amid rude voices? A beloved light?
> A Solitude, a Refuge, a Delight?
>
> (56–64)

In line 59 Shelley uses 'like' to announce a typical transposition; we
are shifted from one figurative representation of Emilia to another.
Phrases such as 'Vanquishing dissonance and gloom' encourage the
reader to think of Emilia in the most general metaphoric terms.
Shelley recreates her in a language which has an almost independent
life. Though Shelley adapts the image of the 'well of sealed and secret
happiness' from the *Song of Songs*,[7] he cuts it off from its sacred
origins. What saves the passage from mere phrase-making, however,
is the poet's attempt to define. The language has the quality of a

----

[7] See Wasserman, 421–2.

litany, but remains alert to nuance of phrase. Envisaged in the line, 'A lovely soul formed to be blest and bless', for example, is a reciprocity of blessing which sustains the idea of relationship. Yet the derealizing, self-defining impact of Shelley's abstractions is apparent. The poetry is always tending to cancel itself, move beyond the separate line or phrase. This is partly the effect of the poem's distinctive prosodic make-up, the impression it gives of combining couplet and blank verse.[8] The readiness of Shelley's phrases to run into each other is one of the poem's most daring and, in places, disquieting tactics. Still, it is possible to measure the linguistic life which *Epipsychidion* does possess by comparing Byron's parallel in *The Island* to these lines at the conclusion of Shelley's poem:[9]

> We shall become the same, we shall be one
> Spirit within two frames, oh! wherefore two?
> One passion in twin-hearts, which grows and grew,
> 'Till like two meteors of expanding flame,
> Those spheres instinct with it become the same . . .
>
> (573–7)

In Canto II of *The Island* Byron talks of

> The other better self, whose joy or woe
> Is more than ours; the all-absorbing flame
> Which, kindled by another, grows the same . . .
>
> (377–9)

What Byron has lost are crucial words like 'Till' and 'become', and the tension generated by repetition (in Shelley's lines, for example, 'We shall become the same' moves to 'become the same'—not quite a shift of tense, but a fraction closer to the desired goal of union; the seemingly otiose 'grows and grew' performs a similar function). In Shelley these details take the strain as the poetry nerves itself for the subsequent address. Heralding the poem's final 'collapse', the friction between the restful cadence of 'the same' and the onward

---

[8] George Saintsbury speaks of the poem as 'attaining a rhymed verse-paragraph which, quite unlike *Lycidas* in particular effect, resembles it in belonging to the general class of "rhymed blank verse"—rhymed verse that acquires the powers of blank, and blank verse that borrows the attraction of rhyme', *A History of English Prosody: From the Twelfth Century to the Present Day* (3 vols., London, 1906–10), iii. 111.

[9] Byron's debt in *The Island* to *Epipsychidion* is suggested in Charles E. Robinson, *Shelley and Byron: The Snake and Eagle Wreathed in Fight*, 239.

push of 'expanding' and 'become' is characteristic of *Epipsychidion*. Byron's 'other better self' gestures lazily at the idea of union intricately explored by Shelley.

Epipsychidion is at its greatest when it persuades the reader that it is apprehending a reality beyond language, as it does in the lines following line 91. By contrast, the passage which immediately precedes it seems to apprehend only the reality of its own language. It is easier (at times too easy) to suggest the 'unapprehended relations of things' by relying on abstractions which can be coaxed into giving up their usual meaning. At the start of this paragraph, Shelley remodels the idea of 'Death', presenting it as an escape from constraints:

> She met me, Stranger, upon life's rough way,
> And lured me towards sweet Death; as Night by Day,
> Winter by Spring, or Sorrow by swift Hope,
> Led into light, life, peace.
>
> (72-5)

There is something arch about the turn to the reader, the 'Stranger', even though the subsequent passage amply justifies Shelley's switch of focus—the switch from 'you' to 'her' giving a more public dimension to the poet's praise. Worrying, too, is the way the lines refuse to be fully serious yet advance symbolic equations which point instantly at the 'unapprehended'. Such doubts, however, are offset by the skill with which Shelley weaves an elaborately self-aware texture:

> An antelope,
> In the suspended impulse of its lightness,
> Were less ethereally light: the brightness
> Of her divinest presence trembles through
> Her limbs, as underneath a cloud of dew
> Embodied in the windless Heaven of June
> Amid the splendour-winged stars, the Moon
> Burns, inextinguishably beautiful:
> And from her lips, as from a hyacinth full
> Of honey-dew, a liquid murmur drops,
> Killing the sense with passion; sweet as stops
> Of planetary music heard in trance.
>
> (75-86)

Here the poetry both calls Emilia's 'divinest presence' into being, and through its ravelled, unravelling flow convinces the reader that this 'presence' is 'too deep | For the brief fathom-line of thought or sense' (89–90). The epithet, 'divinest', for instance, hovers between complimenting and redefining. Emilia is later said to be a 'mortal shape indued | With love and life and light and deity' (112–13), but even then 'deity' has a humanist rather than transcendental colouring.

The verse falls tactfully short of 'Killing the sense with passion', being more intent on paving the way for what is to come. It makes the reader aware of something 'inextinguishably beautiful' which is on the verge of manifesting itself. It dwells on each stage of a fuller coming-into-focus. This lingering has a reflexive element: the poem comes into being as the woman does. The language is full of surprises which awaken acceptance of paradox. There are the confusions of synaesthesia—Emilia's lips let fall 'a liquid murmur'; there is the use of 'Embodied' to describe the cloud that—as a figurative equivalent for Emilia's 'limbs'—has just disembodied the woman; there is the stress-shift across the line-ending of 'the Moon | Burns'. The lines adroitly refuse to allow priority to a 'spiritual' or a 'physical' understanding of Emilia. She is known only through a process of veilings and unveilings; she exists sensuously and supra-sensuously.

But *Epipsychidion* is at its finest when, inspired by a subject (Emilia's 'being') beyond language, it rises to this:

> The glory of her being, issuing thence,
> Stains the dead, blank, cold air with a warm shade
> Of unentangled intermixture, made
> By Love, of light and motion: one intense
> Diffusion, one serene Omnipresence,
> Whose flowing outlines mingle in their flowing,
> Around her cheeks and utmost fingers glowing
> With the unintermitted blood, which there
> Quivers, (as in a fleece of snow-like air
> The crimson pulse of living morning quiver,)
> Continuously prolonged, and ending never,
> Till they are lost, and in that Beauty furled
> Which penetrates and clasps and fills the world;
> Scarce visible from extreme loveliness.
> Warm fragrance seems to fall from her light dress,
> And her loose hair; and where some heavy tress
> The air of her own speed has disentwined,
> The sweetness seems to satiate the faint wind;

And in the soul a wild odour is felt,
Beyond the sense, like fiery dews that melt
Into the bosom of a frozen bud.—

(91–111)

Celebrating both 'The glory of her being' and their own sustained poetic life, the couplets participate memorably in what Shelley calls 'The air of her own speed'. But the poetry's life is the more passionate for the reader's sense of its potential transience. It flickers and cross-flickers in the space between a Platonic absolute ('that Beauty') and the particular person. The poetry is 'visionary' in its hints of epiphany, its fidelity to the unseen, its use of language to transform and transfigure what at the outset is presented as 'actual', 'real'. Here the poetry's terms are its own. It is impossible to disentangle that 'warm shade | Of unentangled intermixture, made | By Love, of light and motion'. Familiar abstractions lose their separateness, coming together in an original 'intermixture'. This is not to deny precision and particularity to the language; as Timothy Webb points out, the 'passage seems to combine the language of science with that of religious devotion'.[10] But the two languages blend and intermix, each giving up its implicit claim to offer a definitive way of apprehending reality; they are used by the poet in the attempt to give individual and passionate expression to his sense of Emilia's being. Thus, the paradoxical verb, 'Stains', pays tribute to the ungraspable source of a glory known only as an enriching 'staining'. The writing is intricate, subtle, intense.

Moreover, this passage is no isolated achievement. It owes part of its power to the way it echoes and resolves earlier motifs. For instance, Emilia's 'glory' emerges from Shelley's previous assertion: 'All shapes look glorious which thou gazest on' (32). When Shelley describes the vision's effect on the beholder, he develops the image of 'dew' (used at 79), but this time makes it 'fiery' (110), a wholly unphysical emblem. For a fleeting moment, the poetry is not engaged in 'Killing the sense', but takes us 'Beyond the sense'. Though Panthea's vision in *Prometheus Unbound*, Act IV ends with a collapse—'and like my words they were no more' (318)—it is far less precarious than this passage in *Epipsychidion*, which begins to

---

[10] *Percy Bysshe Shelley: Selected Poems*, ed. Timothy Webb, 217.

lose momentum as Emilia and that Being merge. The poetry's attempt
to resist its loss of momentum is poignant:

> See where she stands! a mortal shape indued
> With love and life and light and deity,
> And motion which may change but cannot die;
> An image of some bright Eternity . . .
>
> (112–15)

The initial cry announces its own latent futility. The greatness of
the poetry is bound up with the way it has made Emilia very difficult
to see. The visual has been gathered up within the visionary.[11] As
Wasserman points out, lines 112–13 identify Emilia with Christ.[12]
Yet the strain of this identification can be felt as Shelley tries to invest
a 'mortal shape' with his own version of 'deity'. And to redefine the
'motion' of line 94 as a quality which 'may change but cannot die'
is ominously wish-fulfilling. The poetry heads inexorably towards
the collapse of line 123 ('Ah, woe is me!'). It is, however, a collapse
that knows the solace of a 'joyous truth' (127). The flow and ebb
of inspiration is charted artfully yet naturally.

   Not until the final section of the poem does Shelley rekindle the
sense that now, in this poem, reality is being transformed. For
example, this stabilizing of poetic energy—

> We—are we not formed, as notes of music are,
> For one another, though dissimilar;
> Such difference without discord, as can make
> Those sweetest sounds, in which all spirits shake
> As trembling leaves in a continuous air?
>
> (142–6)

—is delicately written. Yet 'continuous', as it steadies 'trembling',
leans heavily on the earlier vision's 'flowing outlines . . . | Continuously
prolonged, and ending never'. The very stability of the image in these
later lines suggests decoration, retreat.

   The sophisticated defence of 'True Love' (160), drawn in part from
Dante's *Purgatorio* (xv. 46–75), should be seen as an ecstatic escape
from the difficulty of sustaining a visionary poetry, not unlike the
reflections which follow *The Prelude*'s 'spots of time'. The writing

---

[11] See Harold Bloom's fine reading of this passage in 'Visionary Cinema of
Romantic Poetry', *The Ringers in the Tower*, esp. 51.
[12] Wasserman, 425.

is accomplished, as these lines on the affinity between Love and Imagination demonstrate, even if the doctrinal note offers a less demanding experience than the visionary:

> Love is like understanding, that grows bright,
> Gazing on many truths; 'tis like thy light,
> Imagination! which from earth and sky,
> And from the depths of human phantasy,
> As from a thousand prisms and mirrors, fills
> The Universe with glorious beams, and kills
> Error, the worm, with many a sun-like arrow
> Of its reverberated lightning.

(162–9)

Shelley makes an apposite link across the poem between Imagination which draws its light 'As from a thousand prisms and mirrors' and Emilia's influence, a 'Mirror | In whom, as in the splendour of the Sun, | All shapes look glorious which thou gazest on'. Such a link is characteristic of the poem's highly but provisionally organized structure.

The poem proper, preceded and concluded by a short introduction and envoi, falls into three sections: lines 1–189 containing rhapsodic invocation and self-conscious recognition of inadequacy, deliberately staged collapse and theoretical rallying; lines 190–387, a long passage of retrospective spiritual autobiography, couched in symbol; and lines 388 to the end, consisting mainly of the poet's invitation to Emilia to accompany him to, and live with him on, an island paradise. Earl Wasserman argues that the poem obeys a drive to organize and reorganize itself in the following way: 'In the first movement, nearly all the essential images and themes of the poem have been deposited in disarray, and the succeeding stages will be successive efforts to draw upon that pool of elements and organize them into a stable, meaningful, and satisfying pattern of relationships.'[13] Jean Hall sees the poem as less piloted by a search for truth than Wasserman does. She insists, rather, that the poet 'shows us ideal episodes of self-development, transformations in which the human strives towards the divine'.[14] Her emphasis on the poet's interest in subjective transformations is welcome. For her, too, however, the poem reveals a coherent pattern of development and growth, the second section

---

[13] Wasserman, 430.
[14] *The Transforming Image: A Study of Shelley's Major Poetry* (Urbana, Chicago, and London, 1980), 103.

offering symbolic transformations of the 'banalities' of the opening with its 'conventional complementary [*sic*] formulas of love poetry'.[15] What should be stressed is that the poetry's moods, modes, and degrees of success vary continually. Shelley keeps pushing beyond any resting-point. The famous attack (147–159) on 'modern morals' (154) and the defence of 'True Love' discussed above both follow on and fall away from the achievement of 'The glory of her being' passage. The contrast between these lines, attacking blinkered fidelity, and two lines from Blake's *Visions of the Daughters of Albion* is suggestive. Shelley writes:

> Narrow
> The heart that loves, the brain that contemplates,
> The life that wears, the spirit that creates
> One object, and one form, and builds thereby
> A sepulchre for its eternity.

> (169–73)

whereas Blake gives us this:

> Such is self-love that envies all, a creeping skeleton
> With lamplike eyes watching around the frozen marriage bed.

> (7: 21–2)

For all its brio, Shelley's language wilts into phrase-making beside Blake's arresting image. What is surprising and exciting about *Epipsychidion*, however, is the number of switches of tone it makes, accommodating a variety of ways of feeling and thinking, different intensities of imagining. The paragraph attacking 'that great sect' (149) is full of fine polemical scorn. The paragraph praising 'True Love' is never less than thought-provoking. And the subsequent passage of imaginative autobiography emerges quite easily, yet takes the reader by surprise.

This autobiographical section (190–387) is often thought to be riddled with weaknesses, the writing lapsing, in Harold Bloom's phrase, from 'confrontation to history'[16] and full of self-pity. Certainly, the passages where Shelley describes in the present tense his sense of what Emilia is are the best things in the poem: they put language and feeling under greatest pressure; they create rather than remember; they warrant, as already suggested, the title 'visionary', dealing as they do in transformations of reality that lie 'too deep | For

[15] *The Transforming Image*, 114.   [16] *SM* 213.

the brief fathom-line of thought or sense'. But the section of the poem devoted to history (what Shelley calls 'an idealized history of my life and feelings')[17] is also necessary and, in its own way, very achieved. Shelley needs this section to put the significance of Emilia into emotional context. It is achieved in its intensity and scope, though it is, perhaps, short on self-criticism. Its main strength, though, is its sense of the importance of the emotional life. Shelley takes risks here—risks of posturing and tactlessness—as he treats his own life as an example, if not as exemplary. Passages such as these waver between self-regard and a kind of bravura:

> Then, from the caverns of my dreamy youth
> I sprang, as one sandalled with plumes of fire,
> And towards the loadstar of my one desire,
> I flitted, like a dizzy moth, whose flight
> Is as a dead leaf's in the owlet light,
> When it would seek in Hesper's setting sphere
> A radiant death, a fiery sepulchre,
> As if it were a lamp of earthly flame.—

(217–24)

> In many mortal forms I rashly sought
> The shadow of that idol of my thought.

(267–8)

> Then, as a hunted deer that could not flee,
> I turned upon my thoughts, and stood at bay,
> Wounded and weak and panting . . .

(272–4)

If such lines are guilty of posturing, it is posturing of a high order. The first passage evokes and playfully judges idealist questing—to soar clear in its penultimate line of self-deprecation. Shelley makes us aware, for a brief moment, that in his earlier questing he had found merely a 'fiery sepulchre', not the more benignant 'fiery dews' (110) to which he has already compared the effect of Emilia's presence. In the second passage Shelley's use of 'rashly' does not imply that he was wrong, but concedes that he was foolishly impetuous, a hint of self-blame which is set against the next line. There, the tormenting elusiveness of the object of the poet's search is caught in the words; he searches for something or someone out there who is only a

---

[17] *Letters*, ii. 434.

'shadow' of an 'idol' (inward image) of his thought. The third passage is less defensible, more obviously moved by itself. Yet what is fascinating about all three passages is their re-entrance into the past, the poet's compulsive reliving of his own experience.

Self-pity for Shelley, as for other poets, is not entirely disastrous. In this section it sparks off a process of self-exploration and self-discovery which takes on larger importance. Shelley is writing generally about the fate of an idealist, both as a way of evading while hinting at unspeakable emotional particulars and as a way of liberating himself from the merely subjective. The writing is, to use Shelley's phrasing from the Preface to *Alastor*, 'allegorical of one of the most interesting situations of the human mind'. But the poetry inhabits a zone where allegory, for all the effect of incidental phrases such as 'the wintry forest of our life' (249) with its Dantesque colouring, is hardly distinguishable from unparaphrasable symbol. If anything, we may have cause to criticize the poem for sealing itself off too much from the emotional realities to which it seems to refer.

The major weakness of this section—indeed of the poem as a whole—is not that Shelley has written in a way that requires autobiographical decoding, but that its intuitions of the 'before unapprehended relations of things' can assume a formulaic character. An example is the equation which Shelley tries to make between the 'self' and the 'world'. As Wasserman puts it, the poem is propelled by the 'drive . . . to transform relationship into identity'.[18] When Shelley makes this theme explicit, though, it is difficult not to see the poem as imprisoned within a purely verbal universe:

> But neither prayer nor verse could dissipate
> The night which closed on her; nor uncreate
> That world within this Chaos, mine and me,
> Of which she was the veiled Divinity,
> The world I say of thoughts that worshipped her . . .
>
> (241–5)

Although there is precise intelligence here—the poetry preserves boundaries even as it seeks to dissolve them—Shelley has taken to spelling out a meaning (as in that insistent last line) which his language and rhythms fail to spark into life. This kind of self-consciousness lacks the challenge of his earlier concentration on the

---

[18] Wasserman, 447.

poetic process. Metaphors and comparisons establish their own reality too easily; the ingenuity of 'the air-like waves | Of wonder-level dream' (195–6) is facile. The following passage is neatly organized rather than intricately developing:

> At length, into the obscure Forest came
> The Vision I had sought through grief and shame.
> Athwart that wintry wilderness of thorns
> Flashed from her motion splendour like the Morn's,
> And from her presence life was radiated
> Through the grey earth and branches bare and dead;
> So that her way was paved, and roofed above
> With flowers as soft as thoughts of budding love;
> And music from her respiration spread
> Like light . . .
>
> (321–30)

The central idea at work in these lines—'And from her presence life was radiated'—curls parasitically round the earlier vision of 'The glory of her being, issuing thence'. At times in this paragraph (321–44) describing the successful culmination of the poet's search, the language ceases to grapple with particulars in order to transform them; rather, it takes for granted that such a transformation has already taken place. The lines above convert experience into a kind of figurative algebra. In the following example, simile plaits together abstractions too unstrenuously and offers untestable equations:

> Soft as an Incarnation of the Sun,
> When light is changed to love, this glorious One
> Floated into the cavern where I lay . . .
>
> (335–7)

Nevertheless, this long central section contains much that is more troubled and more exploratory. Often the writing occupies a space between the very private and the boldly confessional. It swings back and forth between wanting to confide and striving to withhold. This can seem coy, even tasteless, as in the lines, 308–16, which many have taken to refer to Harriet Shelley. Yet it can lead to the expression of charged states of feeling. Shelley's very obliquity can create haunting effects:

> One stood on my path who seemed
> As like the glorious shape which I had dreamed,
> As is the Moon, whose changes ever run
> Into themselves, to the eternal Sun;
> The cold chaste Moon, the Queen of Heaven's bright isles,
> Who makes all beautiful on which she smiles,
> That wandering shrine of soft yet icy flame
> Which ever is transformed, yet still the same,
> And warms not but illumines. Young and fair
> As the descended Spirit of that sphere,
> She hid me, as the Moon may hide the night
> From its own darkness, until all was bright
> Between the Heaven and Earth of my calm mind,
> And, as a cloud charioted by the wind,
> She led me to a cave in that wild place,
> And sate beside me, with her downward face
> Illumining my slumbers, like the Moon
> Waxing and waning o'er Endymion.
> And I was laid asleep, spirit and limb,
> And all my being became bright or dim
> As the Moon's image in a summer sea,
> According as she smiled or frowned on me;
> And there I lay, within a chaste cold bed;
> Alas, I then was nor alive nor dead . . .

(277–300)

The flow of the narrative is crucial; it guides the reader with an eerie logic through a critical stage in a relationship from the poet's initial sense of 'Deliverance' (277) to his almost hypnotized state of being 'laid asleep'. Rephrasings and repetitions are important, enacting the slide from one phase to another. As the word 'Moon' is repeated, it takes on quietly sinister suggestions. In lines 287–8, the unnatural clarity of the simile (a 'she', already compared to the moon, is again compared to the moon) restores the full strangeness of the image. Moreover, the second mention of the moon is preceded by epithets—'cold chaste'—which are conventional enough for the moon, but which already imply sexual difficulties in the poet's relationship. This implication is deepened in the subsequent repetition (with variation) in 'within a chaste cold bed'; it is as if something that the poet had always sensed subliminally had finally been actualized. Such writing gets beyond the posturing or tactless (though it must have been difficult for Shelley to write and harrowing for

Mary Shelley to read). It finds a symbolic idiom through which it can evoke and explore. Shelley is at once recollecting poet and suffering man. His emotional past is transformed because it is put in a context which, provisionally, makes sense of it.

What the concluding section of the poem does is to recover that sense of creating a visionary experience occurring earlier in it. This sense is sharpened by the reader's awareness that Shelley's attack on dualisms will collide ultimately with his desire to write a love poem to another person. Only at the end of the poem does he admit this incompatibility, but the friction it leads to is evident throughout. The reader glimpses a potentially dangerous aspect to Shelley's feelings: the possibility of 'annihilation' (587). The word's meaning is suspended between two senses: one is the quasi-mystical 'annihilation' of self through union with another; the other, more darkly shaded, is the suggestion that total union (between the poet and Emilia) might be destructive. However—especially by contrast with the use of the past tense in the backwardly glancing central section—the final section is optimistic in that it journeys constantly towards a transformed state.

Epipsychidion's previously 'unapprehended' intuitions are most fully developed in the systematic lyricism of the concluding 'Invitation au Voyage'. 'Redefinitions' follow one another in orderly, musical procession:

> And every motion, odour, beam, and tone,
> With that deep music is in unison:
> Which is a soul within the soul—they seem
> Like echoes of an antenatal dream.—
>
> (453–6)

The position of these lines represents an important stage; the poetry has moved imperceptibly beyond description into a symbolic version of a state of wholeness. Without the preceding particulars of 'fountain, rivulet, and pond, | As clear as elemental diamond' (436–7), the writing would seem too abstract; without the ensuing quickening and intensifying, it would risk complacency. At moments such as the mention of 'deep music', the poem grows uncannily aware of itself, as if poetry embodied that principle of order and beauty which this poem pursues.

Yet such self-awareness is not all-engulfing. It is not surprising that an account of the poem published in the 1970s should concentrate on the self-reflexive suggestions of the following lines:

But the chief marvel of the wilderness
Is a lone dwelling, built by whom or how
None of the rustic island-people know:
'Tis not a tower of strength, though with its height
It overtops the woods; but, for delight,
Some wise and tender Ocean-King, ere crime
Had been invented, in the world's young prime,
Reared it, a wonder of that simple time,
An envy of the isles, a pleasure-house
Made sacred to his sister and his spouse.
It scarce seems now a wreck of human art,
But, as it were Titanic; in the heart
Of Earth having assumed its form, then grown
Out of the mountains, from the living stone,
Lifting itself in caverns light and high . . .
And, day and night, aloof, from the high towers
And terraces, the Earth and Ocean seem
To sleep in one another's arms, and dream
Of waves, flowers, clouds, woods, rocks, and all that we
Read in their smiles, and call reality.

(483–97, 508–12)

For J. Hillis Miller the writing undoes its desire to 'reconcile once
more, in a performative embrace, nature, supernature, and man'.[19]
The very language Shelley uses, he argues, evokes 'irreconcilable
compartments separated by the dividing textured membrane which
tries to bring them together'.[20] It is true that Shelley's language
makes us aware of the difficulty of a return to origins. The whole
passage admits the fictionalizing that is going on: 'It scarce seems
now a wreck of human art, | But, *as it were* Titanic' (my italics).
Yet the poetry is not just self-absorbed in the way favoured by
deconstructionists. It stretches beyond itself, not merely inviting the
reader, in William Keach's words, to regard the description as 'a
visualizing of what Shelley's own verse must be like as it is shaped
by and gives shape to the forces of accident, fragmentation, fading
and erasure'.[21] The reference in line 492 to the Ocean-King's 'sister
and his spouse' still holds open the possibility of an escape from
transgression. To take the opposing view involves swimming against
the current of the poetry; it is far too gloomy to say that 'the poet's

---

[19] 'The Critic as Host', *DC* 241.
[20] 'The Critic as Host', 242.   [21] Keach, 153.

attempt to repeat with Emily the pleasure of the Ocean-King and his sister only repeats the crime of illicit sexual relations'.[22] Hillis Miller seems determined to smuggle back the moralizing reality-principle which the writing itself holds at bay. But the passage makes the reader aware of two forces: Shelley's sense that a primal state of innocence is hard to recover and his ability to imagine a future state of innocence that admits and repudiates its belatedness. The quality of play, the impression we get of the poet as a conjurer calling up and dissolving reality, helps here; there is a sophisticated awareness of normal ways of experiencing as fictions: 'Earth and Ocean *seem* | To sleep in one another's arms'; the poet speaks of 'all that we | *Read* in their smiles, and call reality' (my italics).

Certainly, however, the poetry of this final section is richly conscious of its own beauty. In the passage where the island is 'elaborately humanized',[23] Shelley both concedes and ignores the presence of fictionalizing:

> And from the sea there rise, and from the sky
> There fall, clear exhalations, soft and bright,
> Veil after veil, each hiding some delight,
> Which Sun or Moon or zephyr draw aside,
> Till the isle's beauty, like a naked bride
> Glowing at once with love and loveliness,
> Blushes and trembles at its own excess . . .
>
> (470–6)

The writing 'concedes' the presence of fictionalizing through its use of simile and through images which bring about a revelation of the 'isle's beauty'. In evoking this revelation it lays claim to offer more than fictionalizing. Yet the air of climax is only momentary. Perfectly miming an 'intense | Diffusion' of energies, of interchangings of identity, this passage is typical of the final section. It might seem self-indulgent were it not for a covert discontent. The poem moves beyond the isle/bride identification to a spiritualizing of the island: 'Yet, like a buried lamp, a Soul no less | Burns in the heart of this delicious isle' (477–8). In the previous image of an unveiling of delights, Shelley anticipates his celebration of poetry in *A Defence of Poetry*: 'Veil after veil may be undrawn, and the inmost naked beauty of the meaning never exposed.' (*PP* 500.) Here he claims that

---

[22] 'The Critic as Host', 242.   [23] Wasserman, 449.

the 'inmost naked beauty' *can* be exposed, though this claim is undercut in turn by the poem's development.

The final paragraph rises to the challenge which Shelley's language has been approaching all along: to drive beyond reciprocity and achieve union. Often the cumulative effect of his identifications justifies Walter Bagehot's praise: 'In the wildest of ecstasies his self-anatomising intellect is equal to itself.'[24] As Shelley's 'awakening spirit' seeks to escape its own knowledge of 'strange | Distinctions' ('Fragments Connected with *Epipsychidion*', 165 and 166-7, *PW* 429), there is a fiercer movement towards resolution. This issues in a poetry which is peculiarly Shelleyan; everything depends on the larger structure to which the ever-regrouping details surrender themselves. So the fiction of the two lovers 'Possessing and possest by all that is' (549) tries to take shelter within its own space:

> Possessing and possest by all that is
> Within that calm circumference of bliss,
> And by each other, till to love and live
> Be one . . .
>
> (549-52)

Yet the poem strives beyond closure. It is the energy latent in 'till' which the last lines unleash. Though Shelley relies on the topos of ineffability, the poem approaches wordlessness with vigilance:

> And we will talk, until thought's melody
> Become too sweet for utterance, and it die
> In words, to live again in looks, which dart
> With thrilling tone into the voiceless heart . . .
>
> (560-3)

Each swerve of emotion is caught and accommodated by the couplets. As Shelley reaches the wordless climax of his poem, he keeps the flame of its verbal intelligence flickering. Browning's equivalent of these lines in *Pauline*—'our hearts so beat together | That speech seemed mockery' (866-7)—has none of the transforming thrust of Shelley's language. Nothing in *Pauline* corresponds to Shelley's ending where affirmation struggles to survive in the rarefied conditions it has created for itself. The pitch of the final cry—

---

[24] 'Percy Bysshe Shelley', *Literary Studies* (1911; 2 vols., London and Toronto, 1916), i. 111.

The winged words on which my soul would pierce
Into the height of love's rare Universe,
Are chains of lead around its flight of fire.—
I pant, I sink, I tremble, I expire!

(588–91)

—though extreme, has been fully, even exhaustively, 'earned'. To
say that it cheats us into talk of the 'limits of poetry' is to suggest
something of the power and importance of *Epipsychidion*. Poignantly,
indeterminately, the impact of the ending wavers: you can see it as
saying either that words cannot render adequately the poet's intuition
of rapture, or as implying that the verbal breakdown corresponds
to the poet's realization of separateness. I incline towards the
second view, a view which might be glossed by this sentence from
Italo Calvino: 'At the moment when you most appear to be a united
voi, a second person plural, you are two *tu*'s, more separate and
circumscribed than before.'[25]

---

[25] *If on a Winter's Night a Traveller*, trans. William Weaver (London, 1982), 123.

8

The Triumph of Life
Questioning and Imagining

It has been an underlying contention of this study that Shelley is a poet who speaks most compellingly when voicing, with a subtle swiftness peculiarly his own, the interplay between aspiration and despair. No poem he wrote is more attuned to this interplay than *The Triumph of Life*. It may imitate the stanza form of Dante's *Divine Comedy*, but exiled from the assured world-view of the earlier poem, it offers an exploration of reality which questions and imagines in its own terms.[1]

To ask questions is for Shelley a means of mobilizing the imagination, of spotlighting areas of shadow. 'For what are we? Whence do we come, and whither do we go?' (*PP* 476.) This threefold enquiry in the essay *On Life* beats at the heart of *The Triumph of Life*, quickens it into poetic existence. But how do the forces of questioning and imagining work in the poem? In what ways do they shape our experience of *The Triumph of Life*?

Hazlitt whets the reader's sense of the poem's destructive edge when he writes: 'The poem entitled the *Triumph of Life*, is in fact a new and terrific *Dance of Death*; but it is thus Mr Shelley transposes the appellations of the commonest things, and subsists only in the violence of contrast.'[2] The poem may seem to subvert the optimistic values observable (in however unstable a form) in Shelley's previous work. Yet its 'violence of contrast' is not the whole story. Animated

---

[1] Paul de Man in 'Shelley Disfigured' also draws attention to 'the questions that articulate one of the text's main structures', *DC* 39, while Angela Leighton points out that the poem's questions (specifically those at lines 176–9) 'reveal the poet's wish to find an answering voice in the landscape', Leighton, 164. But de Man is led to ask questions about the whole purpose of asking questions about a text that asks questions; my reading is less 'metacritical'. Leighton's concern with Shelley's treatment of the sublime also provides a more thematic and theoretical emphasis than I attempt to offer; my stress is laid on the imaginative achievement of the poem.

[2] Review of *Posthumous Poems*, in Redpath, *The Young Romantics and Critical Opinion*, 393.

by Shelley's discontent with human limitations, the poem has as little time for pessimism as for optimism as a fixed stance. *The Triumph of Life* neither rejects nor reaffirms the poet's Promethean faith in the human spirit, a faith which, as Chapter Five in particular has argued, is itself rarely embraced complacently by Shelley. Instead, the poem tests such faith.

Something of the nature of this 'testing' and of the imaginative means which Shelley uses can be gathered by setting the poem alongside Coleridge's *The Rime of the Ancient Mariner*. Both poems draw sustenance from genres—ballad, dream-vision—which allow the two poets to externalize their deepest fears and desires. Both swing between the clear-cut and the inexplicable. In both, consciousness altering, alters all. In Coleridge's poem, as in Shelley's, the reader reaches states of mind through attending to images: the sinister beauty of the ice mast-high, the shadow surrounding the becalmed ship, the unattainable serenity of the climbing moon. Both poets describe a condition on earth which could be called purgatorial. Coleridge stresses the disappearance of God and the mystery of evil. The Mariner's actions, whether good or bad, exclude motive and transcend awareness. Shelley emphasizes the paradox that 'Good' and 'the means of good' are 'irreconcilable', the difficulty, at times the seeming impossibility, of sustaining ideals of love, hope, and relationship.

Of course, there are differences. Shelley's sense of reality is more involuted, though no more frightening than Coleridge's. *The Triumph of Life* appends no moral, nor does it seem likely to have done so had it been finished. It does not concentrate on one consciousness, but two. Yet in its visionary power, its narrative flow, its symbolic explorations, its obsession with the Sun, its use of simile, and its stress on subjectivity, *The Triumph of Life* has fascinating affinities with *The Rime of the Ancient Mariner*. In his poem, Shelley is always approaching, though never quite reaching, that trust in the symbol which shines through Coleridge's famous account:

a Symbol . . . is characterized by a translucence of the Special in the Individual or of the General in the Especial or of the Universal in the General. . . . It always partakes of the Reality which it renders intelligible; and while it enunciates the whole, abides itself as a living part in that Unity, of which it is the representative.[3]

---

[3] *The Collected Works of Samuel Taylor Coleridge*, vi. *Lay Sermons* (1972), ed. R. J. White (London and Princeton, NJ, 1969–), 30.

If Coleridge sees symbols as incarnating meaning, Shelley's practice in *The Triumph of Life* is to pursue meaning through images. The difference can be felt in the imaginative impact of the two poems. Coleridge deals in nightmare, Shelley in phantasmagoria. *The Ancient Mariner* settles in the mind as a series of glowing oil-paintings. *The Triumph of Life* affects one as a process unfolding at speed; the lines one has read seem like the dazzling wake left by some tireless, onward-moving energy.

What this chapter will study is how, in *The Triumph of Life*, imagining entails questioning and questioning leads to further imagining, how power and elusiveness coexist, whether this coexistence warrants praise or blame. As Richard Cronin points out, F. R. Leavis's description of the poem is, as description, accurate: 'The poem itself is a drifting phantasmagoria—bewildering and bewildered. Vision opens into vision, dream unfolds within dream, and the visionary perspectives . . . shift elusively and are lost.'[4] Cronin argues that *The Triumph of Life*'s 'value as a poem depends very much on how far its reader is prepared to accept and be satisfied with its labyrinthine inconclusiveness'.[5] This 'inconclusiveness' is presented lucidly, more the correlative of that drive in Shelley which made him allegedly say, 'I always go on until I am stopped, and I never am stopped'[6] than of a mandarin, self-delighting scepticism.

In support of these claims for *The Triumph of Life*, one might consider the passage where the poem's 'I' (referred to in this chapter as 'the Poet') looks round to see who gave the answer 'Life' to the questions which escaped him while reflecting on the contents of his terrifying vision:

> Struck to the heart by this sad pageantry,
> Half to myself I said, 'And what is this?
> Whose shape is that within the car? & why'—
>
> I would have added—'is all here amiss?'
> But a voice answered . . 'Life' . . . I turned and knew
> (O Heaven have mercy on such wretchedness!)
>
> That what I thought was an old root which grew
> To strange distortion out of the hill side
> Was indeed one of that deluded crew,

---

[4] Leavis's account quoted in Cronin, 221–2.   [5] Cronin, 222.
[6] Quoted in Edward John Trelawny, *Records of Shelley, Byron, and the Author*, ed. David Wright (Harmondsworth, 1973), 111.

And that the grass which methought hung so wide
  And white, was but his thin discoloured hair,
And that the holes it vainly sought to hide

  Were or had been eyes.—'If thou canst forbear
To join the dance, which I had well forborne,'
  Said the grim Feature, of my thought aware,

'I will tell all that which to this deep scorn
  Led me and my companions, and relate
The progress of the pageant since the morn;

  'If thirst of knowledge doth not thus abate,
Follow it even to the night, but I
  Am weary' . . . Then like one who with the weight

Of his own words is staggered, wearily
  He paused, and ere he could resume, I cried,
'First who art thou?' . . . 'Before thy memory

  'I feared, loved, hated, suffered, did, and died,
And if the spark with which Heaven lit my spirit
  Earth had with purer nutriment supplied

'Corruption would not now thus much inherit
  Of what was once Rousseau—nor this disguise
Stain that within which still disdains to wear it.—'

                        (176–205)

Perhaps the first thing that strikes one about this is the inventiveness of the writing, the imagining that is going on. Eliot quoted the lines as proof of Dante's *'cumulative* influence'[7] on Shelley, and indeed they are evidence that 'the greatest poetry can be written with the greatest economy of words'.[8] But it is a very Shelleyan form of economy seen at work here and elsewhere in the poem, a verbal tautness that is arresting in its details, fluid in its overall sweep. This passage's rhythms and syntax keep pace with the Poet's dawning horror, while its images capture Rousseau's degradation as a state that the reader can see and, after the reference to the 'holes it vainly sought to hide', feel sympathy for. Questioning liberates the writing, arousing a desire to explore riddles of experience. The Poet's enquiry—'who art thou?'—prompts Rousseau to plunge into self-accusing, self-excusing autobiography. This warms the passage to

  [7] 'What Dante Means to Me', *To Criticize the Critic* (1965; London, 1978), 130.
  [8] T. S. Eliot, 'Dante', *Selected Essays* (3rd enlarged edn., 1951; London, 1976), 252.

life, and makes it much more than a rhetorical *tour de force*. Rousseau is not just the object of ambivalent opinion as he is in Byron's *Childe Harold*, Canto III, stanzas LXXVII–LXXXI, or in this verdict from Mary Wollstonecraft's *A Vindication of the Rights of Woman*: 'had Rousseau mounted one step higher in his investigation, or could his eye have pierced through the foggy atmosphere, which he almost disdained to breathe, his active mind would have darted forward to contemplate the perfection of man in the establishment of true civilisation, instead of taking his ferocious flight back to the night of sensual ignorance'.[9] Less interested in judging the 'real' Rousseau than in using him to explore his own concerns, Shelley animates his creation. For all his emphasis on historical contextualizing, Edward Duffy would seem to agree with this position when he asserts: 'Shelley's Rousseau tells his life Shelley's way, not his own.' Yet Duffy attempts to bridge 'Shelley's way' and 'his [Rousseau's] own' when he goes on to say: 'Like the historical Rousseau, the one whom Shelley creates speaks from the conviction that for the most telling record of public events one can do no better than to turn to one's own private register of experience.'[10] Shelley was obviously fascinated by Rousseau, but Duffy's argument slights the transformation of historical figure into imaginative creation which the poem carries out. We can hear Shelley's Rousseau. The poetry lets Rousseau's account weigh heavily with us: does it endorse his version of what has happened to him? The full answer to 'who are thou?' involves consideration of the Poet's earlier unspoken question, ' "& why"— | I would have added—"is all here amiss?" ' This in turn means that the reader must await Rousseau's central narrative, which tells of an encounter with a 'shape all light' (352). So, for all their power, Rousseau's words are the tip of an iceberg: the poem confronts us with a submerged bulk of issues which have yet to be resolved. Throughout *The Triumph of Life*, Shelley refuses to paraphrase his poem's complex experience, or yield that experience to the mind's abstracting ambitions. Although there are attempts in the poem to make sense of life in conceptual terms, these attempts mark stages rather than goals.

---

[9] Ed. Miriam Brody Kramnick (Harmondsworth, 1975), 99.
[10] *Rousseau in England: The Context for Shelley's Critique of the Enlightenment* (Berkeley, Los Angeles, and London, 1979), 107.

Comparison with a twentieth-century heir of the poem, Wilfred Owen's 'Strange Meeting', may help to bring this last point into sharper focus. The later poem compacts and miniaturizes the Romantic dream-vision. While its title echoes a phrase in Shelley's *The Revolt of Islam* (V. 1831, *PW* 82), its manner has more in common with *The Triumph of Life* or with Keats's *The Fall of Hyperion*. It arrests these poems at their point of highest intensity, confrontation with the 'Other'. But it parcels itself into two halves: the visionary and the admonitory. The vision is clear and powerful, hauntingly rendering the speaker's entrance into chthonic halls and his encounter with the dead soldier. The poetry undergoes a slowing of pulse:

> With a thousand pains that vision's face was grained;
> Yet no blood reached there from the upper ground,
> And no guns thumped, or down the flues made
>   moan.[11]

The admonitions are, by turns, profound and teasingly difficult; they are sometimes, one may feel, both at once:

> Now men will go content with what we spoiled,
> Or, discontent, boil bloody, and be spilled.

The poem waives continuous quest in favour of wisdom. Shelley, by contrast, sustains vision into and beyond admonition. There is in Owen, as in the encounter in Eliot's 'Little Gidding', a suggestion of catharsis, a purgation of terror achieved by summoning up, and hearing the gnomic words of, a 'familiar compound ghost'.[12] It is in no way a diminishing of their poetic achievement to say that Owen and Eliot, by using vision as a vehicle for taking a full look at the worst, provide a kind of consolation which Shelley denies the reader of *The Triumph of Life*.

The Triumph of Life and 'Strange Meeting' are visionary reports on history, which helps both to account for their strengths and to suggest the dangers they court. Though their generalizations often resonate, they can also obscure, and neither poem is free from hints of phrase-making. More often, however, Shelley's language chisels its meanings even when paradox is its theme:

---

[11] Quoted from *The Collected Poems of Wilfred Owen*, ed. C. Day Lewis (1963; London, 1971).
[12] Quoted from *The Complete Poems and Plays of T. S. Eliot* (1969; London, 1973).

I felt my cheek
Alter to see the great form pass away
Whose grasp had left the giant world so weak
That every pigmy kicked it as it lay—
And much I grieved to think how power and will
In opposition rule our mortal day—
And why God made irreconcilable
Good and the means of good . . .

(224-31)

What is impressive about these lines is the ease with which generalization rises out of the visionary narrative (the account of Napoleon's fall) and is absorbed back into it. The Poet's despairing insight is finely prepared for, flowing from a carefully evoked state of 'alteration': 'I felt my cheek | Alter . . . '. Again, the poetry fuses the clear, the disturbing, and the mysterious. Shelley does, and does not, give the Poet's analysis his imprimatur. The passage bears the stress of a decade's thinking about the clash between 'Good' and 'the means of good';[13] it voices the despair which the Fury employs as a temptation in *Prometheus Unbound* (I. 618-31). Yet we are also asked to read the moment as dramatic, exposing the Poet's need for certainty. As Rousseau implies later (248-51),[14] the Poet's stance of detached sorrow ignores his own unavoidable involvement in historical process. Shelley may have recourse to 'God', the most absolute of answers. But—and this contributes towards its achievement—the passage sparks off a swarm of unspoken, onward-propelling questions.

What Shelley says in *The Triumph of Life* does not, for all its power, necessarily have the support of a cut-and-dried moral perspective. There is much in the poem to buttress the view that it indicts all those who fail an 'absolute humanitarian standard'.[15] Yet, time and again, the poem makes distinctions only to blur their relevance as explanations. Often an enriching quality in this blurring

[13] For a personal application of this insight, see Shelley's letter to Mary Shelley: 'good far more than evil impulses—love far more than hatred—has been to me, except as you have been it's object, the source of all sort[s] of mischief', *Letters*, ii. 339.
[14] A rejected passage shows Rousseau more explicitly warning the Poet against the dangers of despair. See App. C, *TLA*-'B', Donald H. Reiman, *Shelley's 'The Triumph of Life': A Critical Study Based on a Text Newly Edited from the Bodleian Manuscript* (Urbana, Ill. 1965), 240-2.
[15] Cameron, 460.

of moral responses makes itself felt, as when Rousseau offers the
following explanation of why he fell:

> I was overcome
> By my own heart alone, which neither age
> Nor tears nor infamy nor now the tomb
> Could temper to its object.

(240–3)

Entangling as they clarify, these lines fascinate; they have the thrust
of a verdict and the strut of a boast. The syntax deflects attention
from what did overcome Rousseau by concentrating on what did
not, while the laying of blame on 'my own heart alone' comes close
to impasse. If the phrase judges Romantic confessionalism, it brings
to its judgement a mixture of exhaustion and respect.

The poem's momentum can knock down edifices that judgement
constructs. The grim description of the dance round the chariot is
an example:

> Wilder as it grows,
> They, tortured by the agonizing pleasure,
> Convulsed and on the rapid whirlwinds spun
> Of that fierce spirit, whose unholy leisure
> Was soothed by mischief since the world begun,
> Throw back their heads and loose their streaming hair . . .
> and as they glow
> Like moths by light attracted and repelled,
> Oft to new bright destruction come and go,
> Till like two clouds into one vale impelled
> That shake the mountains when their lightnings mingle
> And die in rain,—the fiery band which held
> Their natures, snaps . . . ere the shock cease to tingle
> One falls and then another in the path
> Senseless, nor is the desolation single,
> Yet ere I can say *where* the chariot hath
> Past over them; nor other trace I find
> But as of foam after the Ocean's wrath
> Is spent upon the desert shore.

(142–7, 152–64)

More than an indictment of lust, these lines suggest an exhilarating,
imaginative release from what feel like extreme presures. Shelley

presents human relationships as if they conformed to impersonal laws of destructiveness. The intensity of feeling in 'agonizing pleasure' strains beyond the immediate sexual overtones. In the same way, the rhythm of 'ere the shock cease to tingle | One falls and then another in the path | Senseless, nor is the desolation single' parades its virtuosity, lingering over the 'desolation', almost relishing the state. Shelley prises open lines from *Adonais*, 'and what still is dear | Attracts to crush, repels to make thee wither' (473–4), and offers, instead, 'and as they glow | Like moths by light attracted and repelled, | Oft to new bright destruction come and go'. The simile depersonalizes and extends the introspective glimpse offered by the lines from *Adonais*. Picking its way round a series of short words, line 154 imparts powerful emphasis to 'destruction', yet displays reserve on Shelley's part. He demands both that the reader should respond to the line's hammer-blow and respect the poet's almost scornful aloofness from the effects his language contrives. The whole passage illustrates how Shelley's imaginative energies successfully break any categorizing mould. Peremptory, wilful, brilliant, the poetry fully expresses its vision of destructiveness. But *The Triumph of Life* puts despair into a context which is at once profounder and more disquieting. Shelley's poem is no simple palinode to Romantic ideals. That Life has conquered its victims is indisputable. Why this should be is answered by judgements that are at once absolute and provisional. Recurring questions respond to this twinning of clarity and enigma. Along with the parallelisms in the narrative and the patterning of symbols, they attempt to map a complex terrain.

Two passages will serve to exemplify the imaginings which create this terrain. The first deals with the onset of 'vision'; the second with its content. The first is the Proem (1–40); the second the Poet's vision of the Car of Life (41–106). Both passages reveal an impressive vitality. *The Triumph of Life* begins with *élan*. When the poem's 'I' is introduced, the effect is to bring into contact power and elusiveness. Shelley is able to convey the strangeness of the experience without having, as in *Prometheus Unbound*, to hold up the narrative,[16] and he avoids the dogged overstatement employed by Byron in *Cain*: 'Thou speak'st to me of things which long have swum | In visions through my thought' (I. i. 164–5). Instead, his negative ('Which

---

[16] See exchange between Earth and Prometheus at I. 107–52.

was not slumber') captures the reality of vision, cunningly refusing to pin it down:

> When a strange trance over my fancy grew
>   Which was not slumber, for the shade it spread
> Was so transparent that the scene came through
>   As clear as when a veil of light is drawn
> O'er evening hills they glimmer; and I knew
>   That I had felt the freshness of that dawn,
> Bathed in the same cold dew my brow and hair
>   And sate as thus upon that slope of lawn
> Under the self same bough . . .
>
> (29–37)

Shelley takes his reader into the depths of the mind, recreating one of its uncanniest states, *déjà vu*. Where the opening lines celebrate the natural, these use it to intimate, compellingly but delicately, the mind's experience. So the Poet's trance is described as having the transparency of a veil of light through which a line of hills is seen to glimmer. The language itself brightens and glimmers with the promise of some revelation. Yet the reader is drawn into the Poet's condition—a state of apartness as touching as any in Romantic poetry—by way of under-emphasis, the mingling of the confessional and the withheld in 'thoughts which must remain untold' (21). A fuller measure of this under-emphasis can be taken by studying Shelley's revision in lines 26–8 ('before me fled | The night; behind me rose the day; the Deep | Was at my feet, and Heaven above my head') of lines from Goethe's *Faust* which Philip Wayne translates as follows:

> The streams of quenchless light I long to drink,
> Before me day and, far behind, the night,
> The heavens above me, and the waves below:
> A lovely dream, but gone with set of sun.[17]

Shelley's Poet sets his back against the natural world with which Faust is despairingly in love. Where Faust diagnoses, Shelley leaves unspoken: a reticence which catches the breath.

---

[17] *Faust, Part One*, trans. Philip Wayne (1949; Harmondsworth, 1975), 66.

The first part of the opening reveals a different kind of reticence. On rereading, nearly all the words that carry emphasis—'hastening', 'splendour', 'mask', 'awakened', 'tempered'—buckle with suspect possibilities:

> Swift as a spirit hastening to his task
>   Of glory and of good, the Sun sprang forth
> Rejoicing in his splendour, and the mask
>
>   Of darkness fell from the awakened Earth.
> The smokeless altars of the mountain snows
>   Flamed above crimson clouds, and at the birth
>
> Of light, the Ocean's orison arose
>   To which the birds tempered their matin lay.
>
>   (1–8)

The poem which begins with the Sun springing forth so that the 'mask | Of darkness fell from the awakened Earth' will break off with an unmasking that exposes an absence of essence rather than a veiled inner being: 'Mask after mask fell from the countenance | And form of all' (536–7). And yet the poetry—visibly enjoying itself—is full of virtuosities, such as the rhythmic change at the start of the last line of this passage:

> All flowers in field or forest which unclose
>   Their trembling eyelids to the kiss of day,
> Swinging their censers in the element,
>   With orient incense lit by the new ray
>
> Burned slow and inconsumably . . .
>
>   (9–13)

The opening, then, stops a fraction short of making the decisive contrast between the natural and the human which Shelley drew in one of his final letters: 'But Nature is here as vivid and joyous as we are dismal, and we have built, as Faust says, "our little world in the great world of all" as a contrast rather than a copy of that divine example.'[18] For all the Proem's echoes of a Miltonic paradisal harmony,[19] it is content to adumbrate, rather than lay claim to, such a meaning. The lines are the result of much labour, and one motive behind Shelley's revisions is the desire to clamp down on

---

[18] *Letters*, ii. 415.
[19] Compare Shelley's opening with *Paradise Lost*, ix. 192–9.

pointers towards symbolic meaning. In one of the discarded openings the Sun is seen as working rather like the Spirit's 'plastic stress' in *Adonais* (381). It is described as burning through its mortal mask 'until the world it treads assume | The beauty of the presence which makes bright | Its desarts from the cradle to the tomb'.[20] What we are given is a passage that excludes ways of suggesting meaning that Shelley has used in previous works, a passage in which meaning inheres in rhythm and image. It establishes being before it implies value. So, the surprise of lines 19–20, referring to 'the toil which he [i.e. the Sun] of old | Took as his own and then imposed on them,' slides into the poem, catching us off guard. For—and this is the key point—however much the critic may see the Sun as the symbolic villain of the poem, nothing can obliterate the fact of its beauty and splendour at the start. If it initiates the central metaphoric action of the poem—'a sun's extinction of a star'[21]—it enjoys, as many passages in the poem do, brief but complete freedom from the significance which symbol supplies.[22]

What is surprising about the next section of the poem, up to and including the appearance of the Car of Life, is an 'ease and grace . . . born of effort which stops short of strain'.[23] Invention, here, avoids the fanciful or strained, and has the fluidity one would expect of a 'waking dream' (42). The poetry defines purposelessness, its subject, through statements which imply questions the crowd ignores but provokes:

> yet none seemed to know
> Whither he went, or whence he came, or why
> He made one of the multitude . . .
>
> (47–9)

To suggest the crowd's collective anonymity, Shelley uses images—such as 'Numerous as gnats upon the evening gleam' (46)—which have the flavour of attenuated epic simile. Although Canto III of

[20] App. B, *DO*–'B', 4–6, Reiman, *Shelley's 'The Triumph of Life'*, 235.
[21] Cronin, 218.
[22] Harold Bloom would defer, but hope ultimately to uncover, what the sun 'signifies' (his word). For him, the sun is 'an emblem of major importance in the poem' (*SM* 224); for me, what is of major importance is that it is impossible to treat the sun as an emblem at any stage in the poem.
[23] James Rieger, *The Mutiny Within: The Heresies of Percy Bysshe Shelley* (New York, 1967), 220.

Dante's *Inferno* has influenced this passage,[24] Shelley chastens the tragic reverberations of Dante's 'great refusal'.[25] But the reader may feel that the detached irony evident in 'serious folly' (73) is not the most intense note in the passage. There is the swift deadliness with which a vicious circle is described in lines 54–5: 'Some flying from the thing they feared and some | Seeking the object of another's fear'. And the satiric tone latent here is more apparent in lines which dissect attitudes to death:

> And others as with steps towards the tomb
> Pored on the trodden worms that crawled beneath,
> And others mournfully within the gloom
>
> Of their own shadow walked, and called it death . . .
> And some fled from it as it were a ghost,
> Half fainting in the affliction of vain breath.
>
> (56–61)

That last line, exposing the absurdity of those who make an 'affliction' out of their expenditure of breath, demonstrates the severity of Shelley's visionary satire.

The Triumph of Life lifts itself into greatness in the following lines where the poetry makes its impact less through isolated felicities than as a whole:

> And a cold glare, intenser than the noon
> But icy cold, obscured with [   ] light
> The Sun as he the stars. Like the young Moon
>
> When on the sunlit limits of the night
> Her white shell trembles amid crimson air
> And whilst the sleeping tempest gathers might
>
> Doth, as a herald of its coming, bear
> The ghost of her dead Mother, whose dim form
> Bends in dark ether from her infant's chair,
>
> So came a chariot on the silent storm
> Of its own rushing splendour, and a Shape
> So sate within as one whom years deform

[24] Cf. Canto III, 67–9, and line 57 of *The Triumph of Life*, as well as Canto III, 112–14, and line 51 of *The Triumph of Life*. See Reiman, *Shelley's 'The Triumph of Life'*, 27–8.
[25] *Inferno*, iii. 60.

Beneath a dusky hood and double cape
  Crouching within the shadow of a tomb,
And o'er what seemed the head a cloud like crape

  Was bent, a dun and faint etherial gloom
Tempering the light; upon the chariot's beam
  A Janus-visaged Shadow did assume

The guidance of that wonder-winged team.
  The Shapes which drew it in thick lightnings
Were lost: I heard alone on the air's soft stream

  The music of their ever moving wings.
All the four faces of that charioteer
  Had their eyes banded . . . little profit brings

Speed in the van and blindness in the rear,
  Nor then avail the beams that quench the Sun
Or that these banded eyes could pierce the sphere

  Of all that is, has been, or will be done.—
So ill was the car guided, but it past
  With solemn speed majestically on . . .

(77–106)

These lines compose a single imaginative curve whose sweep begins
with 'cold glare', takes in the subtle but transparent simile of the
'young Moon', introduces the chariot, its occupant, and driver,
before passing, as the chariot passes, 'With solemn speed majestically
on'. The writing is, for the most part, impressively independent of
sources and analogues; it seems to come out of nowhere but the poet's
imagination. *The Triumph of Life* reworks material from writings
by a number of authors, including Ezekiel, Dante, Petrarch, Calderón,
and Wordsworth. Yet what is fascinating is the way so echoic a poem
goes about its business as if its imagining were primary. Though,
for instance, the irony of 'o'er what seemed the head' with its
evocation of Milton's Death is not lost on the reader,[26] it does not
persuade one to interpret Shelley's Shape in terms other than those
proposed by the passage.[27] Directing his vision to the imagination's
eye, Shelley brings out the strangeness of what he sees. The result

---

[26] *Paradise Lost*, ii. 672.
[27] Though her emphasis differs from mine, Miriam Allott is also interested in
what this borrowing suggests about Shelley's 'search for appropriate figurative
language'. See 'The Reworking of a Literary Genre: Shelley's "The Triumph of Life" ',
EOS 270.

is definition of an object by controlled reference to the associations it arouses, a form of definition used in the fine simile starting at line 79, where the interplay of effects is adroitly managed. There is a suggestion of approached and transgressed limits—'the sunlit limits of the night', the old moon in the young moon's arms. There is a suggestion, too, of peculiar beauty, sufficiently unlike that of the chariot arriving on 'the silent storm | Of its own rushing splendour' to prepare us for that splendour (at once bogus and real) without forestalling its impact. At the outset of the passage Shelley's rhythms quicken powerfully in response to the 'cold glare, intenser than the noon | But icy cold'. Swiftly arranging the different lights into a symbolic hierarchy, lines 77–9 discover a meaning in Shelley's vision that the reader can neither challenge nor allegorize.

Similarly, comparisons deepen and riddle one's grasp of the Shape. The reader's imagination is at full stretch, picturing her as inside the chariot and as 'Crouching within the shadow of a tomb'. The line is imported from a more straightforward context in *Adonais*, 'Seek shelter in the shadow of the tomb' (458). Part of the poetry's assault on limits, visual and imaginative, the moment contributes to the passage's visionary chiaroscuro, to a scene of 'gloom | Tempering the light', of 'A Janus-visaged Shadow', of magnificence and ineptness, of wonder and fear. The vision of the chariot is too resonant for solely parodic intonations to win the day. The language is not interested merely in the unmaking of myth, in deflation. Tones modulate from rapture loaded with menace ('I heard alone on the air's soft stream | The music of their ever moving wings') to proverbial forthrightness ('little profit brings | Speed in the van and blindness in the rear'). The fineness of the writing does not lie in the 'violence of contrast' with which the lines repudiate a literary tradition. It derives, rather, from the energy of invention it displays. The final lines—'but it past | With solemn speed majestically on'—end the vision on a fittingly equivocal note. Shelley wants the reader to speak 'majestically' with complete trust in the word's sonority and with full-blooded irony. As Angela Leighton points out, the 'whole presentation of the chariot is ambiguous', offering 'the onlooker a choice of interpretations'.[28]

This episode is crucial to a poetic experience which is both labyrinthine and elegantly ordered. At the heart of *The Triumph of*

[28] Leighton, 161.

Life, intriguingly related to the first vision of a 'Shape', is another vision, an enigma, the 'shape all light' who appears to Rousseau as an ideal of beauty, annihilates his former thoughts and wanes in the light of a 'new Vision' (411) of which she is at once herald and antagonist. The passage describing this encounter begins with an echo of the Poet's *déjà vu* in the opening. But Rousseau's intimations of a previous existence lack such assurance as the earlier moment possessed:

> Whether my life had been before that sleep
> The Heaven which I imagine, or a Hell
>
> Like this harsh world in which I wake to weep,
> I know not. I arose and for a space
> The scene of woods and waters seemed to keep,
>
> Though it was now broad day, a gentle trace
> Of light diviner than the common Sun
> Sheds on the common Earth, but all the place
>
> Was filled with many sounds woven into one
> Oblivious melody, confusing sense
> Amid the gliding waves and shadows dun . . .
>
> (332–42)

His sole guarantee of a previous existence is his sense of a 'gentle trace | Of light diviner than the common Sun | Sheds on the common Earth'. This trace is one which 'The scene of woods and waters' only 'seemed to keep', a 'light diviner' that both modulates and lingers on into 'the Sun's image radiantly intense' (345) from which the shape all light emerges. The blurring of boundaries has as its agent and correlative the 'one | Oblivious melody' which entrances Rousseau. This haunting lyricism is typical of the whole episode. Shelley imagines an encounter which is untranslatable.

Even at the frightening climax, this untranslatable quality is forced on the reader by the involved figurativeness of the writing:

> And still her feet, no less than the sweet tune
> To which they moved, seemed as they moved, to blot
> The thoughts of him who gazed on them, and soon
>
> All that was seemed as if it had been not,
> As if the gazer's mind was strewn beneath
> Her feet like embers, and she, thought by thought,

> Trampled its fires into the dust of death,
> As Day upon the threshold of the east
>   Treads out the lamps of night, until the breath
>
> Of darkness reillumines even the least
>   Of heaven's living eyes—like day she came,
> Making the night a dream . . .

(382–93)

These lines are the crest of a wave of ever-increasing pressure. Yet if 'seemed' has shadowed the shape all light's beauty, it sustains a wavering uncertainty which stops any too speedy judgement of her. Along with 'as if', it is there at the kill, insinuating doubt and trance when the reader may long for certainty and blame. The shape all light's malevolence is unequivocal only for a moment. Shelley does insist that the simile is taken exactly in 'like day she came, | Making the night a dream', and yet he allows his climactic moment to go, implying the provisionality of his simile when the shape all light who extinguished the stars is herself compared to the waning morning star (412–15).

That Rousseau should use figures of speech at the climax of his account is both interesting and crucial. Vision goes beyond its own setting to find terms in which to convey significance, to explain itself to itself. In doing so, it gives an eerie sense of Rousseau's state of trance (he refers to himself in the third person) and admits that significance of a fixed kind eludes its grasp. The structure of the writing is richly intricate, suspending meaning even as it defines it, explaining an experience by a Chinese box-like series of figures. Indeed, an account of a reading experience of the 'shape all light' passage ought to focus less on ambivalence than on the writing's moment by moment approach to and withholding of significance. The move towards identifiable meaning is begun in the following lines:

> As one enamoured is upborne in dream
>   O'er lily-paven lakes mid silver mist
> To wondrous music, so this shape might seem
>
>   Partly to tread the waves with feet which kist
> The dancing foam, partly to glide along
>   The airs that roughened the moist amethyst . . .

(367–72)

Rousseau modifies his account of the shape all light's gliding with a care that is almost feline, the delicate pressure of emphasis in the repeating of 'partly' conveying the strangeness and insubstantiality of the shape all light's presence.

The potential menace of the shape all light is hinted at, too, in the suppressed disclaimer of 'moved in a measure new | Yet sweet' (377–8), lines in which 'Yet' awakens the doubt it seeks to resolve. Doubt emerges as justifiable at the climax (discussed above) and in the similes that follow. Here rival impulses in the language converge as simile, which has receded into its own world of suggestion at several points in the passage, now takes on itself the burden of locating meaning. The result is to rouse the reader's awareness of chiming symbolic suggestions in the poem. But a counter-movement is already beginning, and the insistence that we accept Rousseau's vision on its own terms battles with our desire for certainty. The mingled knowledge and blankness, sureness and doubt haunt in a way that almost oppresses.

The passage from one phase of consciousness into another has been accomplished through a visionary encounter which neither Rousseau nor reader would be without, yet whose value is flickering and whose connection with disenchantment is definite, though susceptible of no straightforward causal explanation. The shape all light enters Rousseau's life at a crucial point—before he wakes to weep; before, that is, disenchantment with its clear sense of loss and confused but aching nostalgias possesses him. It is the passage into that condition which the episode of Rousseau's being offered a cup by the shape all light brings about:

> Shew whence I came, and where I am, and why—
> Pass not away upon the passing stream.
>   'Arise and quench thy thirst,' was her reply.
> And a shut lily, stricken by the wand
>   Of dewy morning's vital alchemy,
> I rose; and, bending at her sweet command,
>   Touched with faint lips the cup she raised,
> And suddenly my brain became as sand
>   Where the first wave had more than half erased
> The track of deer on desert Labrador,
>   Whilst the fierce wolf from which they fled amazed
> Leaves his stamp visibly upon the shore
>   Until the second bursts—so on my sight
> Burst a new Vision never seen before.—
>
>   (398–411)

Shelley's imaginings are at their uncanniest here, and once again a question, this time asked by Rousseau, releases them. The onward impulsion of narrative—hastened by a series of 'ands' which keep an open mind about causal links but communicate a sense of tranced flow—is important. What has to be asked of Shelley's poem, however, is whether it is succumbing to mannerism, giving in to the instinct to evade. Simile follows simile in a procession of figures which halt at the frontier of explicit significance. The image of the 'shut lily' is a case in point. Shelley's language may vividly intimate, but the lines do not stabilize the value of the simile, unlike their probable source in Dante, a passage which C. H. Sisson renders in this way:

> As little flowers, which in a frosty night
> Droop and shut tight, when the sun shines on them
> Stretch and look up, erect upon their stalks,
>
> So I recovered from my failing strength . . .[29]

Whereas the last line in the passage from the *Inferno* completes the circuit of meaning, in Shelley's hands the vehicle of a comparison is often related to its tenor in a resonant but indefinite way. This use of language is defensible, indeed enriching. Shelley's is a rendering of experience which does not pretend not to be a rendering. So here he is able to enact the successive layerings and unlayerings of mental events. The mind whose thoughts have been trampled into extinction undergoes a further loss of identity. Simile erases simile as wave erases wave. With great delicacy (evident in 'more than half'), Shelley renders that state of transition when 'the first wave had more than half erased | The track of deer on desert Labrador'. His images of deer and wolf tantalize with emblematic hints, just as the line, 'And suddenly my brain became as sand', is, for a moment, terrifying before the ensuing tercet softens threat into something warier, stranger. The reader passes into the most poignant stretch of the poem where the shape all light is herself subject to erasure:

> So knew I in that light's severe excess
> The presence of that shape which on the stream
> Moved, as I moved along the wilderness,

---

[29] *Inferno*, ii. 127–30, *The Divine Comedy*, trans. C. H. Sisson (London and Sydney, 1981), 55.

More dimly than a day appearing dream,
  The ghost of a forgotten form of sleep,
A light from Heaven whose half extinguished beam

  Through the sick day in which we wake to weep
Glimmers, forever sought, forever lost.—
  So did that shape its obscure tenour keep

Beside my path, as silent as a ghost;

(424–33)

Rousseau experiences a grief that communicates itself to the rhythms and images. They are weighted with a sense of loss, animated by a despairing ghost of hope. The passage avoids self-indulgence. It understands from within the nature of nostalgia, its obsessive fixation on an image of loss. 'Glimmers' is especially haunting. After the entrancing light Rousseau has seen, all that remains is this residual flicker. Once again, the writing creatively imagines the layering of experience.

To define the passage's unique achievement, one might contrast it with Wordsworth's 'Immortality Ode', a poem which it tantalizingly echoes. For the assurance of significance which underpins Wordsworth's striking transitions—'O joy! that in our embers | Is something that does live, | That nature yet remembers | What was so fugitive!' (133–6)—Shelley substitutes vision with all its doubts, seemings, glimmerings. Only they can explain themselves, and what they explain is an experience imagined in an utterly different way from Wordsworth's. Shelley's is an elusiveness which springs from an unusually sharp awareness of the gap between experience and interpretation, not the tic of a poetry fearful of botching the meanings it has within its grasp. Where Wordsworth's myth of pre-existence structures his light symbolism, Shelley relies on a pure narrative imagining. As a result we get, at the heart of the poem, a vision which refuses to be allegorized, and which insists on the subjectivity and difficulty of ascribing value.

Wordsworth's instincts are different. Whatever the strangeness of the self's experience, the poetic effect is robustly untantalizing. When in *The Prelude* he recalls his experience of 'visionary dreariness', his poetry is every bit as involved with enigma as Shelley's in the 'shape all light' passage. But how differently enigma is presented:

              It was, in truth,
    An ordinary sight, but I should need

Colours and words that are unknown to man
To paint the visionary dreariness
Which, while I looked all round for my lost guide,
Did at that time invest the naked pool,
The beacon on the lonely eminence,
The woman, and her garments vexed and tossed
By the strong wind.[30]

Wordsworth initiates us in the extraordinary by rooting us in the 'ordinary', by calmly asserting the uniqueness of his experience. This experience is reinforced by the charged reintroduction of the objects invested with his feelings: the pool, beacon, woman. Images, but not images for images as in Shelley, they draw to themselves the full weight of Wordsworth's sense of 'visionary dreariness'. Subjectivity is for Wordsworth a sturdier, possibly a stranger business. This should not lead to an undervaluing of Shelley's achievement in the account of the shape all light in *The Triumph of Life*, the strength of which lies in its unappeased passage between possible meanings.

In the two central visions of his poem, Shelley does not gloss his imaginings, or does so only to a limited extent. The second of these visions, Rousseau's meeting with the shape all light, is bounded by two Shelleyan sorrows, both highlighted by questions. First, there is the agonizing over human ills which Rousseau tells the Poet (327–31) he would forget had he heard the 'oblivious spell' (331) of the water, ills that have prompted the Poet's questions to Rousseau. Secondly, there is Rousseau's own inquiry into origins and ends, 'Shew whence I came, and where I am, and why', which is a sorrow because the question goes unanswered, only instigating the next stage of his forgetting. *The Triumph of Life* is unusually 'question-proof' as well as being resistant to the act of faith which elsewhere in Shelley's poetry his scepticism can incite. *Adonais*, for example, obeys a reckless urge to answer these grieving questions:

Whence are we, and why are we? of what scene
The actors or spectators?

(184–5)

---

[30] *The Prelude, 1805*, xi. 307–15. Edn. used is *'The Prelude', 1799, 1805, 1850*, Norton Critical Edition, ed. Jonathan Wordsworth, M. H. Abrams, Stephen Gill (New York and London, 1979).

A clue to the difference between the two poems is offered by the reappearance of that last phrase—significantly modified—in a passage from *The Triumph of Life*. Rousseau, in answer to the Poet's questions, says, 'But follow thou, and from spectator turn | Actor or victim in this wretchedness' (305–6). *The Triumph of Life* commits itself to immersion in the probably destructive element of life. *Adonais*, similarly bleak in its diagnosis of existence, leaves behind 'the contagion of the world's slow stain' (356), answering its questions through an act of faith that admits its willed quality. The affirmations of the final section of *Adonais* both outrage and exhilarate; they move beyond human limitations yet tacitly mourn the cancelled reality of Shelley's more humanist declarations. The elegy concludes in the silence of a mythical world of Being which stands beyond poetry:

> The soul of Adonais, like a star,
> Beacons from the abode where the Eternal are.
>
> (494–5)

The Triumph of Life has more in common with these lines from *Faust*, lines translated by Philip Wayne with a power which helps to explain why Shelley quoted the original appreciatively:[31]

> The spirit's splendour, in the soul unfurled,
> Is ever stifled with a stranger stuff.[32]

But in finding a visionary language to express the spirit's struggle with a 'stranger stuff', Shelley avoids nihilism. His poetry can be credited with awareness of the positive values whose destruction it records. This point is given support by lines towards the end of the fragment:

> I became aware
> Of whence those forms proceeded which thus stained
>   The track in which we moved; after brief space
> From every form the beauty slowly waned,
>
>   From every firmest limb and fairest face
> The strength and freshness fell like dust, and left
>   The action and the shape without the grace

---

[31] *Letters*, ii. 364.   [32] *Faust, Part One*, 52.

Of life; the marble brow of youth was cleft
  With care, and in the eyes where once hope shone
Desire like a lioness bereft

  Of its last cub, glared ere it died; each one
Of that great crowd sent forth incessantly
  These shadows, numerous as the dead leaves blown

In Autumn evening from a poplar tree—
  Each like himself and like each other were,
At first, but soon distorted, seemed to be

  Obscure clouds moulded by the casual air;
And of this stuff the car's creative ray
  Wrought all the busy phantoms that were there

As the sun shapes the clouds . . .

                                        (516–35)

Offering an answer to Dante, a 'wonder worthy of [his] rhyme'
(471), these lines blend power with bitterness and establish themselves
as original. The underscoring of Rousseau's speaking voice—'I
became aware'—sharpens the poetry. Phantasmagoria is presented
with chastened elegance, as in the quickly modelled simile in lines
524–6, or in the purging of Miltonic grandeur, Vallombrosa and
its fallen leaves being reduced to the crisply etched image of the
deciduous poplar tree. The *terza rima* pursues its course as unre-
mittingly as the 'car's creative ray', detonating the ironies that were
latent in the opening lines of the poem. There is equal emphasis on
shaping and insubstantiality. So 'casual' concedes the absence of
control which 'moulded' had seemed to claim. The car's creative ray
foments formlessness. The vitality of the writing—its tribute to 'the
grace | Of life'— is the sole but crucial affirmation to emerge from
the passage, as Hazlitt brings out in his comment on this part of
the poem: 'Any thing more filmy, enigmatical, discontinuous,
unsubstantial than this, we have not seen; nor yet more full of morbid
genius and vivifying soul.'[33]
    It is impossible to say whether the poem would have ended close
to where it does had Shelley lived. Yet the fragmentary form imposed
on *The Triumph of Life* by Shelley's untimely death seems in keeping
with the poem's deepest instincts; certainly in its concluding lines
the fragment drives to the core of the problem it explores, especially

[33] Redpath, *The Young Romantics and Critical Opinion*, 393–4.

in the Poet's words: ' "Then, what is Life?" I said' (544). Would this question have stimulated further imaginings? What one can assert is that Shelley has done what Chekhov defined as the artist's primary duty:[34] he has formulated the questions correctly, doing so through imaginings adequate to his lucid, subtle, and passionate vision of the contradictions and uncertainties that constitute experience.

[34] 'You are right to demand that an author take conscious stock of what he is doing, but you are confusing two concepts: *answering the questions* and *formulating them correctly*. Only the latter is required of an author', *Anton Chekhov's Life and Thought: Selected Letters and Commentary*, trans. Michael Henry Heim in collaboration with Simon Karlinsky (Berkeley, Los Angeles, and London, 1973), 117.

Bibliography

ABBEY, LLOYD, *Destroyer and Preserver: Shelley's Poetic Skepticism* (Lincoln, Nebr., and London, 1979).

AESCHYLUS, *'Prometheus Bound', 'The Suppliants', 'Seven Against Thebes', 'The Persians'*, trans. Philip Vellacott (1961; Harmondsworth, 1970).

ALLOTT, MIRIAM, 'The Reworking of a Literary Genre: Shelley's "The Triumph of Life" ', in Miriam Allott (ed.), *Essays on Shelley (EOS)*, 239-78.

—— (ed.), *Essays on Shelley* (Liverpool, 1982).

BAGEHOT, WALTER, *Literary Studies* (1911; 2 vols., London and Toronto, 1916).

BAKER, CARLOS, *Shelley's Major Poetry: The Fabric of a Vision* (1948; Princeton, NJ, 1973).

BERRYMAN, JOHN, *The Freedom of the Poet* (New York, 1976).

BLAKE, WILLIAM, *The Complete Writings of William Blake*, Oxford Standard Authors, ed. Geoffrey Keynes (London, New York, and Toronto, 1966).

BLOOM, HAROLD, *Shelley's Mythmaking* (1959; Ithaca, NY, 1969).

—— *The Ringers in the Tower: Studies in Romantic Tradition* (Chicago and London, 1971).

—— *A Map of Misreading* (1975; Oxford, New York, Toronto, and Melbourne, 1980).

—— *Poetry and Repression: Revisionism from Blake to Stevens* (New Haven and London, 1976).

—— BLOOM, HAROLD, DE MAN, PAUL, DERRIDA, JACQUES, HARTMAN, GEOFFREY H., MILLER, J. HILLIS, *Deconstruction and Criticism (DC)* (London and Henley, 1979).

BLUNDEN, EDMUND, *Shelley: A Life Story* (1946; London, New York, and Toronto, 1965).

BRISMAN, LESLIE, 'Byron: Troubled Stream from a Pure Source', *ELH* xlii (1975), 623-50.

BRISMAN, SUSAN HAWK, ' "Unsaying His High Language": The Problem of Voice in *Prometheus Unbound*', *SIR* xvi (1977), 51-86.

BROWNING, ROBERT, *Poetical Works: 1833-1864*, Oxford Standard Authors, ed. Ian Jack (London, New York, and Toronto, 1970).

BÜCHNER, GEORG, *The Plays of Georg Büchner*, trans. Victor Price (London, Oxford, and New York, 1971).

BUSH, DOUGLAS, *Mythology and the Romantic Tradition in English Poetry* (1937; New York, 1957).

BUTLER, MARILYN, *Romantics, Rebels and Reactionaries: English Literature and its Background 1760-1830* (Oxford, New York, Toronto, and Melbourne, 1981).

BYRON, Lord, *Poetical Works*, Oxford Standard Authors, ed. Frederick Page, new ed. and corr. John Jump (3rd edn., London, Oxford, and New York, 1970).

CALVINO, ITALO, *If on a Winter's Night a Traveller*, trans. William Weaver (London, 1982).

—— *Mr Palomar*, trans. William Weaver (London, 1986).

CAMERON, KENNETH NEILL, *Shelley: The Golden Years* (Cambridge, Mass., 1974).

CAVE, RICHARD ALLEN, 'Romantic Drama in Performance', in Richard Allen Cave (ed.), *The Romantic Theatre: An International Symposium*, 79-104.

—— (ed.), *The Romantic Theatre: An International Symposium* (Gerrards Cross, Bucks., and Totowa, NJ, 1986).

CHEKHOV, ANTON, *Anton Chekhov's Life and Thought: Selected Letters and Commentary*, trans. Michael Henry Heim in collaboration with Simon Karlinsky (Berkeley, Los Angeles, and London, 1973).

CHERNAIK, JUDITH, *The Lyrics of Shelley* (Cleveland, Ohio, and London, 1972).

—— AND BURNETT, TIMOTHY, 'The Byron and Shelley Notebooks in the Scrope Davies Find', *Review of English Studies*, new series, xxix (1978), 36-49.

COLERIDGE, S. T., *Poetical Works*, Oxford Standard Authors, ed. E. H. Coleridge (1912; London, Oxford, and New York, 1969).

—— *The Complete Poetical Works of Samuel Taylor Coleridge*, ed. E. H. Coleridge (2 vols., Oxford, 1912).

—— *Biographia Literaria: or Biographical Sketches of my Literary Life and Opinions*, ed. George Watson (1956; repr. with additions and corrections, 1965; London, 1971).

—— *Lay Sermons* (1972), ed. R. J. White, vol. vi of *The Collected Works of Samuel Taylor Coleridge* (London and Princeton, NJ, 1969-).

CONRAD, PETER, *The Victorian Treasure-House* (London, 1973).

CRABBE, GEORGE, *'Tales, 1812' and Other Selected Poems*, ed. Howard Mills (Cambridge, 1967).

CRONIN, RICHARD, 'Shelley's Witch of Atlas', *KSJ* xxvi (1977), 88-100.

—— *Shelley's Poetic Thoughts* (London, 1981).

CURRAN, STUART, *Shelley's 'Cenci': Scorpions Ringed with Fire* (Princeton, NJ, 1970).

—— *Shelley's Annus Mirabilis: The Maturing of an Epic Vision* (San Marino, Calif., 1975).

—— 'Shelleyan Drama', in Richard Allen Cave (ed.), *The Romantic Theatre: An International Symposium*, 61-77.

DANTE ALIGHIERI, 'The Banquet' (Il Convito) of Dante Alighieri, trans. Katharine Hillard (London, 1889).
—— La Vita Nuova, trans. Barbara Reynolds (1969; Harmondsworth, 1975).
—— The Divine Comedy, trans. C. H. Sisson (London and Sydney, 1981).
DAVIE, DONALD, Purity of Diction in English Verse (enlarged edn., 1967; London, 1969).
DE MAN, PAUL, 'Shelley Disfigured', DC 39–73.
DONOGHUE, DENIS, 'Keach and Shelley', London Review of Books, 19 Sept. 1985, 12–13.
DONOHUE, JR., JOSEPH W., 'Shelley's Beatrice and the Romantic Concept of Tragic Character', KSJ xvii (1968), 53–73.
DUFFY, EDWARD, Rousseau in England: The Context for Shelley's Critique of the Enlightenment (Berkeley, Los Angeles, and London, 1979).
ELIOT, T. S., The Complete Poems and Plays of T. S. Eliot (1969; London, 1973).
—— Selected Prose of T. S. Eliot, ed. Frank Kermode (London, 1975).
—— Selected Essays (3rd enlarged edn., 1951; London, 1976).
—— To Criticize the Critic (1965; London, 1978).
EMPSON, WILLIAM, Seven Types of Ambiguity (3rd edn., 1961; Harmondsworth, 1973).
EVEREST, KELVIN, 'Shelley's Doubles: An Approach to Julian and Maddalo', in Kelvin Everest (ed.), Shelley Revalued: Essays from the Gregynog Conference, 63–88.
—— (ed.), Shelley Revalued: Essays from the Gregynog Conference (Leicester, 1983).
FLETCHER, ANGUS, Allegory: The Theory of a Symbolic Mode (Ithaca, NY, 1964).
FOUCAULT, MICHEL, The History of Sexuality, i: An Introduction, trans. Robert Hurley (1978; Harmondsworth, 1984).
FOWLER, ROGER (ed.), A Dictionary of Modern Critical Terms (1973; London, Henley, and Boston, 1982).
FREUD, SIGMUND, 'The "Uncanny"', 'An Infantile Neurosis' and Other Works (1955), vol. xvii of The Standard Edition of the Complete Psychological Works of Sigmund Freud, ed. James Strachey, trans. in collaboration with Anna Freud, assisted by Alix Strachey and Alan Tyson (24 vols., London, 1953–74), 217–56.
GIBSON, EVAN K., 'Alastor: A Reinterpretation', PP 545–69, reprinted from PMLA lxii (1947), 1022–45.
GOETHE, JOHANN WOLFGANG VON, Faust, Part One, trans. Philip Wayne (1949; Harmondsworth, 1975).
HALL, JEAN, The Transforming Image: A Study of Shelley's Major Poetry (Urbana, Chicago, and London, 1980).

HALL, SPENCER, 'Power and the Poet: Religious Mythmaking in Shelley's "Hymn to Intellectual Beauty"', *KSJ* xxxii (1983), 123–49.

HARTMAN, GEOFFREY H., *The Unmediated Vision: An Interpretation of Wordsworth, Hopkins, Rilke, and Valéry* (1954; New York, 1966).

HAWTHORNE, NATHANIEL, *The Marble Faun* (New York, Scarborough, Ont., and London, 1961).

HAZLITT, WILLIAM, *The Spirit of the Age: or Contemporary Portraits*, ed. E. D. Mackerness (London and Glasgow, 1969).

HILDEBRAND, WILLIAM H., 'A Look at the Third and Fourth Spirit Songs: *Prometheus Unbound, I*', *KSJ* xx (1971), 87–99.

HOGLE, JERROLD E., 'Metaphor and Metamorphosis in Shelley's "The Witch of Atlas"', *SIR* xix (1980), 327–53.

—— 'Shelley's Fiction: The "Stream of Fate"', *KSJ* xxx (1981), 78–99.

—— 'Shelley's Poetics: The Power as Metaphor', *KSJ* xxxi (1982), 159–97.

—— Review of William Keach's *Shelley's Style*, *KSJ* xxxv (1986), 183–8.

HUGHES, D. J., 'Coherence and Collapse in Shelley, with Particular Reference to *Epipsychidion*', *ELH* xxviii (1961), 260–83.

—— 'Potentiality in *Prometheus Unbound*' [*SIR* ii (1963), 107–26], in R. B. Woodings (ed.), *Shelley: Modern Judgements*, 142–61.

—— 'Kindling and Dwindling: The Poetic Process in Shelley', *KSJ* xiii (1964), 13–28.

KEACH, WILLIAM, *Shelley's Style* (New York and London, 1984).

KEATS, JOHN, *Poetical Works*, Oxford Standard Authors, ed. H. W. Garrod (1956; London, Oxford, and New York, 1970).

KNIGHT, G. WILSON, *The Starlit Dome: Studies in the Poetry of Vision* (1941; London, Oxford, and New York, 1971).

LEAVIS, F. R., *Revaluation: Tradition and Development in English Poetry* (1936; Harmondsworth, 1972).

LEIGHTON, ANGELA, *Shelley and the Sublime: An Interpretation of the Major Poems* (Cambridge, 1984).

MCFARLAND, THOMAS, 'Recent Studies in the Nineteenth Century', *Studies in English Literature: 1500–1900*, xvi (1976), 693–727.

MCGANN, JEROME J., 'A Reply to George Ridenour', one of a number of short articles, 'On Byron', *SIR* xvi (1977), 571–83 (563–94).

MCNIECE, GERALD, 'The Poet as Ironist in "Mont Blanc" and "Hymn to Intellectual Beauty"', *SIR* xiv (1975), 311–36.

MATTHEWS, G. M., 'A Volcano's Voice in Shelley', *ELH* xxiv (1957), 191–228.

—— 'A New Text of Shelley's Scene for *Tasso*', *KSMB* xi (1960), 39–47.

—— '"Julian and Maddalo": The Draft and The Meaning', *Studia Neophilologica*, xxxv (1963), 57–84.

—— (ed.), *Keats: The Critical Heritage* (London, 1971).

MEYERS, JEFFREY, *Painting and the Novel* (Manchester, 1975).

MILLER J. HILLIS, 'The Critic as Host', *DC* 217–53.

MILTON, JOHN, *Poetical Works*, Oxford Standard Authors, ed. Douglas Bush (1966; London and Oxford, 1969).

MUESCHKE, PAUL, and GRIGGS, EARL L., 'Wordsworth as the Prototype of the Poet in Shelley's *Alastor*', *PMLA* xlix (1934), 229–45.

MULHERN, FRANCIS, *The Moment of 'Scrutiny'* (London, 1979).

MURRAY, E. B., 'Mont Blanc's Unfurled Veil', *KSJ* xviii (1969), 39–48.

NELLIST, BRIAN, 'Shelley's Narratives and "The Witch of Atlas"', *EOS* 160–90.

NEWEY, VINCENT, 'The Shelleyan Psycho-Drama: "Julian and Maddalo"', *EOS* 71–104.

NEWLYN, LUCY, 'Shelley's Ambivalence', review of William Keach's *Shelley's Style*, *EIC* xxxvi (1986), 263–8.

O'NEILL, MICHAEL, 'A More Hazardous Exercise: Shelley's Revolutionary Imaginings', forthcoming in J. R. Watson (ed.), 'The French Revolution in English Literature and Art: Special Number', *The Yearbook of English Studies*, xix (1989).

OWEN, WILFRED, *The Collected Poems of Wilfred Owen*, ed. C. Day Lewis (1963; London, 1971).

PEACOCK, THOMAS LOVE, *The Works of Thomas Love Peacock*, The Halliford Edition, ed. H. F. B. Brett-Smith and C. E. Jones (10 vols., London and New York, 1924–34).

PINSKY, ROBERT, *Landor's Poetry* (Chicago and London, 1968).

PITT, VALERIE, one of a number of contributors to 'Reading Shelley', 'The Critical Forum', *EIC* iv (1954), 99–103 (87–103).

PULOS, C. E., *The Deep Truth: A Study of Shelley's Scepticism* (1954; Lincoln, Nebr., 1962).

REDPATH, THEODORE, *The Young Romantics and Critical Opinion, 1807–1824: Poetry of Byron, Shelley, and Keats as Seen by their Contemporary Critics* (London, 1973).

REIMAN, DONALD H., *Shelley's 'The Triumph of Life': A Critical Study based on a Text Newly Edited from the Bodleian Manuscript* (Urbana, Ill., 1965).

—— *Percy Bysshe Shelley* (1969; London and Basingstoke, 1976).

—— 'The Norton Shelley', *KSJ* xxx (1981), 14.

RIEGER, JAMES, *The Mutiny Within: The Heresies of Percy Bysshe Shelley* (New York, 1967).

ROBINSON, CHARLES E., *Shelley and Byron: The Snake and Eagle Wreathed in Fight* (Baltimore and London, 1976).

ROGERS, NEVILLE, *Shelley at Work: A Critical Inquiry* (2nd edn., Oxford, 1967).

ROSS, MARLON B., 'Shelley's Wayward Dream-Poem: The Apprehending Reader in *Prometheus Unbound*', *KSJ* xxxvi (1987), 110–33.

RUSKIN, JOHN, *The Works of John Ruskin*, ed. E. T. Cook and Alexander Wedderburn (39 vols., London, 1903–10).

SAINTSBURY, GEORGE, *A History of English Prosody: From the Twelfth Century to the Present Day* (3 vols., London, 1906-10).

SCHILLER, FRIEDRICH VON, 'Naive and Sentimental Poetry' and 'On the Sublime', trans. Julius A. Elias (1966; New York, 1975).

SCRIVENER, MICHAEL HENRY, *Radical Shelley: The Philosophical Anarchism and Utopian Thought of Percy Bysshe Shelley* (Princeton, NJ, 1982).

SHAKESPEARE, WILLIAM, *The Complete Works*, ed. Peter Alexander (1951; London and Glasgow, 1973).

SHELLEY, PERCY BYSSHE, *The Complete Works of Percy Bysshe Shelley*, The Julian Edition, ed. Roger Ingpen and Walter E. Peck (1926-30; 10 vols., London, 1965).

—— *Shelley's 'Prometheus Unbound': A Variorum Edition*, ed. Lawrence John Zillman (Seattle, 1959).

—— *Shelley and his Circle 1773-1822*, vols. i-iv, ed. Kenneth Neill Cameron (Cambridge, Mass., and London 1961-70), vols. v-viii, ed. Donald H. Reiman (Cambridge, Mass., 1973-86).

—— *The Letters of Percy Bysshe Shelley*, ed. Frederick L. Jones (2 vols., Oxford, 1964).

—— 'Alastor' and Other Poems; 'Prometheus Unbound' with Other Poems; 'Adonais', ed. P. H. Butter (London and Glasgow, 1970).

—— *Poetical Works*, Oxford Standard Authors, ed. Thomas Hutchinson, corr. G. M. Matthews (2nd edn., London, New York, and Toronto, 1970).

—— *Shelley's Poetry and Prose*, Norton Critical Edition, ed. Donald H. Reiman and Sharon B. Powers (New York and London, 1977).

—— *Selected Poems*, ed. Timothy Webb (London and Totowa, NJ, 1977).

SPENSER, EDMUND, *Poetical Works*, Oxford Standard Authors, ed. J. C. Smith and E. de Selincourt (1912; London, Oxford, and New York, 1970).

SPERRY, STUART M., 'The Ethical Politics of Shelley's *The Cenci*', *SIR* xxv (1986), 411-27.

STEVENS, WALLACE, *The Collected Poems of Wallace Stevens* (1955; London, 1971).

TETREAULT, RONALD, *The Poetry of Life: Shelley and Literary Form* (Toronto, Buffalo, and London, 1987).

TOMLINSON, CHARLES, 'The Poet as Translator', review of Timothy Webb's *The Violet in the Crucible*, *The Times Literary Supplement*, 22 Apr. 1977, 474-5.

TRELAWNY, EDWARD JOHN, *Records of Shelley, Byron, and the Author*, ed. David Wright (Harmondsworth, 1973).

WARD, GEOFFREY, 'Transforming Presence: Poetic Idealism in *Prometheus Unbound* and *Epipsychidion*', *EOS* 191-212.

WASSERMAN, EARL R., *Shelley: A Critical Reading* (Baltimore and London, 1971).

WEBB, TIMOTHY, 'Shelley and the Religion of Joy', *SIR* xv (1976), 357–82.

—— *The Violet in the Crucible: Shelley and Translation* (Oxford, 1976).

—— *Shelley: A Voice Not Understood* (Manchester, 1977).

WEISKEL, THOMAS, *The Romantic Sublime: Studies in the Structure and Psychology of Transcendence* (Baltimore and London, 1976).

WILSON, MILTON, *Shelley's Later Poetry: A Study of his Prophetic Imagination* (New York, 1959).

WOLLSTONECRAFT, MARY, *A Vindication of the Rights of Woman*, ed. Miriam Brody Kramnick (Harmondsworth, 1975).

WOODINGS, R. B. (ed.), *Shelley: Modern Judgements* (London, 1968).

WOODMAN, ROSS, Review of Earl Wasserman's *Shelley: A Critical Reading*, *KSJ* xxi–xxii (1972–3), 244–52.

WORDSWORTH, JONATHAN, 'The Secret Strength of Things', *The Wordsworth Circle*, xviii (1987), 99–107.

WORDSWORTH, WILLIAM, *Poetical Works*, Oxford Standard Authors, ed. Thomas Hutchinson, new ed. Ernest de Selincourt (1936; London, Oxford, and New York, 1969).

—— *'The Prelude', 1799, 1805, 1850*, Norton Critical Edition, ed. Jonathan Wordsworth, M. H. Abrams, Stephen Gill (New York and London, 1979).

WORTON, MICHAEL, 'Speech and Silence in *The Cenci*', *EOS* 105–24.

YEATS, W. B., *A Vision* (1937; London, 1981).

—— *The Collected Poems of W. B. Yeats* (2nd edn., 1950; London and Basingstoke, 1971).

Index

(Figures in bold type indicate extended discussion of a work by Shelley.)

language *(cont.)*:
 and the self 14, 22, 73–91 *passim*,
  157, 158, 160–1, 162–3, 173
 self-awareness in the use of 6–7,
  27, 40, 102, 119, 127, 128–9,
  158, 163–4, 167, 173–5
 self-divided activity of 4
 Shelley on the 'real language of
  men' 78–9
 and social class 52–3, 58
 as theme 74–5, 76–7, 79–80, 88,
  97, 125, 158–9, 176–7
 and virtuosity 142, 159, 160, 186,
  188
 see also conflict, fictions, imagery,
  reflexiveness, tension, and visionary
  poetry
Leavis, F. R. 1, 6, 76 n.11, 180
 on Shelley's 'confusion' 40–1
 on Shelley's poetry as 'peculiarly
  emotional' 30, 31, 35
 see also emotion and philosophy
Leighton, Angela 33, 33 n.8,
  130 n.6, 178 n.1, 192

McFarland, Thomas 1–2
McGann, Jerome J. 144
McNiece, Gerald 40
Matthews, G. M. 9 n.26, 105–6
metaphor: *see* imagery
Meyers, Jeffrey 87 n.24
Miller, J. Hillis 173–5
Milton, John 23, 98
 Lycidas 162 n.8
 Paradise Lost 59, 120, 122, 188,
  191, 200
Mueschke, Paul: and Griggs, Earl L.
  13 n.7
Mulhern, Francis 30
Murray, E. B. 48
myth 92
 attempted construction of 3, 95,
  102
 mythical world of Being 199
 and myth-making 2, 121, 140–2,
  145, 146–7
 in Wordsworth 197

narrative 11–29 *passim*, 179, 197
 onward impulsion of 196
 parallelisms in 186
 and relation to lyric 21, 30
 see also allegory

Nature 23, 26, 28, 29, 41–3
 in contrast with the human 64, 188
 invoked by Spenser 135
 and the mind's experience 187
 participation in and distance from
  55–6
 Shelleyan and Wordsworthian views
  of contrasted 14–15, 136–7
Nellist, Brian 95, 136
Newey, Vincent 63–4, 68, 73
Newlyn, Lucy 7–8

Owen, Wilfred 183

Peacock, Thomas Love 29
Petrarch 159, 191
philosophy 3, 4, 12
 and poetry 32, 40, 43–6
 see also emotion, Platonism, and
  scepticism
Pinsky, Robert 57, 58
Pitt, Valerie 6
Platonism 94, 116, 149, 165
politics 3, 8, 13 n.7, 34, 46, 98 n.15
 involvement in historical process
  184
 millennial poetry 112, 118
 visionary politics and imagery
  105–7
 visionary reports on history 183
Power 42, 45, 86
 attribute of 'Spirit' 38
 equivocal status of 28–9, 50
 unknowability of 35
Pulos, C. E. 12

questioning: *see* language

reflexiveness 5, 17, 19, 123, 129,
  132 n.8, 164, 173
 defined 5 n.18
 delight in 131, 150–2
Reiman, Donald H. 160
Reynolds, Barbara 159 n.3
Rieger, James 189
Robinson, Charles E. 33 n.8,
  162 n.9
Rogers, Neville 160 n.5
Ross, Marlon B. 92–3
Rousseau, Jean-Jacques: and the
  fictional Rousseau of *The
  Triumph of Life* 182
Ruskin, John 103